P9-CLS-755

Also by
Sharon Bertsch McGrayne

The Theory That Would Not Die

Prometheans in the Lab

Nobel Prize Women in Science

Also by
Rita Colwell

Methods in Aquatic Microbiology
(with M. Zambruski)

Estuarine Microbial Ecology
(with L. H. Stevenson)

Effect of the Ocean Environment on Microbial Activities
(with R. Morita)

The Role of Culture Collections in the Era of Molecular Biology

Vibrios in the Environment

Biotechnology in the Marine Sciences
(with E. Pariser and A. Sinskey)

Biotechnology of Marine Polysaccharides
(with E. R. Pariser and A. J. Sinskey)

Method for Coding Data on Microbial Strains for Computers
(with Morrison Rogosa and Micah I. Krichevsky)

Current Methods for Classification and Identification of Microorganisms
(with R. Grigorova)

Biomolecular Data: A Resource in Transition

Microbial Diversity and Ecosystem Function
(with D. Allsopp and D. L. Hawksworth)

The Global Challenge of Marine Biotechnology
(with R. A. Zilinskas, D. A. Lipton, and R. T. Hill)

Microbial Diversity in Time and Space
(with K. Ohwada and U. Simidu)

Nonculturable Microorganisms in the Environment
(with D. J. Grimes)

Oceans and Health: Pathogens in the Marine Environment
(with Shimshon Belkin)

A Lab of One's Own

One Woman's Personal Journey Through Sexism in Science

Rita Colwell, PhD,

and

Sharon Bertsch McGrayne

Simon & Schuster

New York London Toronto Sydney New Delhi

Simon & Schuster
1230 Avenue of the Americas
New York, NY 10020

First Simon & Schuster hardcover edition August 2020

Interior design by Lewelin Polanco

Manufactured in the United States of America

1 3 5 7 9 10 8 6 4 2

Library of Congress Cataloging-in-Publication Data has been applied for.

ISBN 978-1-5011-8127-6
ISBN 978-1-5011-8128-3 (ebook)

To Jack Colwell:
champion golfer, brilliant polymath,
accomplished yachtsman, loving father,
and beloved husband, without whom this book
could never have been written nor my life
have been so blessed and joyful

And to my husband, George F. Bertsch,
without whom this book
could not have been written

Contents

A Note to the Reader

What follows is Rita Colwell's story, told in her voice. But the stories of others, who had similar experiences, are based on interviews conducted by Dr. Colwell and/or Sharon Bertsch McGrayne. To make it easy for the reader, those too are told in the voice of Dr. Colwell.

prologue

Hidden No More

Graduate student Margaret Walsh Rossiter made a habit of attending Friday afternoon beer parties with Yale University's eminent historians of science. One day, out of curiosity, she asked the great men present, "Were there any women scientists?" This was 1969, and none had been mentioned in her courses or reading material.

"No," came the answer. "There have never been any."

"Not even Madame Curie," someone asked, "who won two Nobel Prizes?"

"No. Never. None," was the response. Marie Curie was a drudge who stirred pitchblende for her husband's experiments. According to some of the world's leading male academics, we women scientists did not exist.

A few years later, Rossiter, still curious, found herself thumbing through a biographical encyclopedia titled *American Men of Science*. Despite the name, she discovered that it included entries on more than a hundred women. Rossiter tried to get an academic job to study more women scientists, but no university was interested. And she couldn't get a grant to do her research independently, because no one else knew enough about women scientists to judge her proposal.

Rossiter didn't have much money, but, liberating her parents' second car, a highly unfashionable Dodge sedan, she spent months driving at top speed, crisscrossing the Northeast from the archives of one women's college to another. Then she expanded her search to the rest of the country, trawling through boxes of records in library basements and attic filing cabinets, finding evidence of women scientists everywhere. A representative denounced her on the floor of Congress, arguing that writing about women scientists was a waste of taxpayers' money. The resulting publicity helped even more people learn about her mission, and soon Rossiter was planning a book—although one Harvard professor joked, "That'll be a really short book, won't it?" A dozen publishers brushed off her proposal because everyone "knew" women scientists didn't exist.

Nevertheless, in 1982, the first book of Rossiter's three-volume history, *Women Scientists in America*, began documenting the existence of our hitherto invisible world. Suddenly, reading those pages, we women in science knew we were not alone. We were the intellectual descendants of a long line of women who'd done significant work. As for Rossiter, she expanded the world of science, founded a new area of study, won a MacArthur Foundation "Genius Grant," and became a chaired professor at Cornell University.

• •

As the story of my life as a scientist, this book tells the human side of this history. It tells what it's like for a woman to go into a field so dominated by men that women were rendered invisible. It's about an enterprise in which, even today, many men and women believe the ability to do high-level science is coded by the Y chromosome; in which men are seen as more competent than identically qualified women; in which the more decorated a male scientist is, the fewer women he trains; in which universities hire their junior faculty members from these elite men's labs.

But let me say from the outset: this book is not a litany of complaints. I have had my own laboratory for almost sixty years, and for every man who blocked my way in science, there were six who helped me. Nevertheless, the scientific enterprise remains a deeply conservative institution filled with powerful men—and some women—who reject outsiders, whether women of any stripe, African American men, Latinos, other people of color, immigrants, LGBTQ people, people with disabilities, or anyone else who doesn't fit the stereotype of the white male genius.

Science is an institution struggling to shed its past. And every time I hear someone say, with the best of intentions, that we have to get more women into science, I get irritated. We have *never* had to interest women in science. Everywhere I've looked, there have been hidden figures, working in the shadows of their husbands' labs or in the labs of male allies, in medical museums and libraries, in government agencies, or in low-level teaching positions across the country. There have *always* been highly capable women wanting to be scientists.

But there has also always been a small set of powerful men who wouldn't let women in. Decades later, we still have men who can't believe that they played any role in stopping talented women from following their passion.

So here in this book, I offer some recommendations for what remains to be done to open the doors of opportunity to women scientists—and how women can open those doors for themselves. Because when women speak up despite the forces acting against us, we will succeed. And succeed we must, because the security, economic strength, and social stability—the destiny of every country in the world—depends on us all.

A Lab of One's Own

chapter one

No, Girls Can't Do That!

It's a beautiful spring day in May 1956, and I'm walking across Purdue University's campus with my handsome, six-foot-two fiancé, Jack Colwell, a former GI back from Germany to finish his graduate studies. On our very first date a few weeks earlier, we decided to get married—two months from now. It's been a whirlwind romance! We couldn't know it at the time, of course, but our wedding would mark the start of sixty-two years of happily married life.

Then Professor Henry Koffler sees us.

Henry Koffler is small in stature, but he's a big man on campus, a powerhouse in biology. Even colleagues can feel intimidated by him, especially when he stands close to them to talk. It's not easy for an undergraduate to get time with him, and so, taking advantage of our fortuitous encounter, I tell him—right there on the sidewalk—my good news: I've decided to postpone medical school and do graduate work in bacteriology while Jack finishes his master's degree in chemistry. All I need to make it happen is a fellowship.

"We don't waste fellowships on women," Koffler says, as if telling me an obvious fact of life.

My first reaction is dismay—quickly followed by anger at the injustice of this policy and at his offhandedness in telling me about it. Without financial assistance, there's no way I can continue my studies. But I can't give Koffler the satisfaction of seeing how upset I am. He seems to think I have no future. *Well*, I tell myself, *I will damn well prove you wrong*.

• •

My parents were Italian immigrants. My father, Louis Rossi, was a stonemason and landscape foreman for a construction company in Beverly Farms, Massachusetts. He built tennis courts, swimming pools, sea walls, and even a steeplechase for big waterfront estates north of Boston. The only way he'd been able to get a high school education in Italy was to enter a Roman Catholic seminary and train for the priesthood. When the time came to be ordained, he skipped town, caught a boat to the United States, and, except for baptisms, weddings, and funerals, never set foot in church again. He told us children his Sunday job was cooking dinner. My mother, Luisa DiPalma Rossi, finished elementary school in her small town near Rome but was then forced to abandon her education to work in her aunt's candle shop. She married my father in Italy and then joined him in the United States a few years after his arrival. It was my mother who took us to church.

During the Great Depression, my father saved $3,000 in cash in a pillowcase under my parents' bed and had the audacity to buy a three-bedroom house in a Yankee neighborhood with a good school. Beverly was a bayside town settled by English colonists in 1626, but by the early twentieth century, many Italian immigrants had found work in the local construction and shoe manufacturing industries. This was an uncomfortable time to be Italian in America. A federal report some years before had warned that "certain kinds of criminality are inherent in the Italian race," while the popular weekly magazine the *Saturday Evening Post* editorialized, "If America doesn't keep out the queer, alien, mongrelized people of Southern and Eastern Europe, her crop of citizens will eventually be dwarfed and mongrelized in turn." Not long after my father arrived in America, the Immigration Act of 1924 barred any more Italians from entering the country, and the year before I was born, a poll of white male Princeton University students listed Italians as the nation's third most objectionable ethnic group, after Muslim Turks and African Americans.

And so the evening our big Catholic family moved into our new house in pleasant, Yankee Beverly Cove, there came a knock at the front door. My father answered it to find a city councilman, who introduced himself and said he represented our new neighbors. They had signed a petition pledging to repay my father for his down payment—provided we moved away. "I bought this house," my father replied. "It is fully paid for." Then he closed the door. It was in this house at 113 Corning Street, Beverly, Massachusetts, that I, Rita Barbara Rossi, was born three years later on November 23, 1934.

My parents would have a total of eight children, including a girl who died in the 1918 influenza epidemic and a boy born just before me who left my mother with postpartum depression and who died of pneumonia before he was two. I was the fifth child to survive. After my younger brother and I entered school, our mother went to work in a local factory.

My mother and father were committed parents, but they lived in a time of traditional attitudes about girls. After school, my sisters and I had to stay indoors making beds and doing housework while the boys did chores outside. My brothers argued a lot, and the unfairness of not being heard over their ruckus rankled me. Although I wasn't fond of my dolls' carriage, I didn't think they had the right to disassemble it—without my permission—to make a go-cart and then refuse to let me play with it or with their Lincoln Logs. Nor did it help that when we visited relatives for Sunday dinners, my brothers received gifts like feather headdresses and toy tomahawks while my sisters and I were expected to wash the dishes and clear up. By the time I was five, I already knew in my heart of hearts that one day I would find a way to escape. *I can't complain now*, I promised myself, *but I'm not going to stay here forever*.

In my innocence as I grew up, I thought our neighbors were upper-middle class. Only later did I realize that they were mostly working class: a police officer, a city clerk, a caretaker for a large estate, a janitor who took care of a public school's furnace and whose wife called him "The Engineer." Most couples had only two or three children, and the wives never spoke to my mother or invited her to their parties. It wasn't only my mother they disapproved of. My father raised flowers in our front yard and fruit trees, vegetables, chickens, and rabbits in the back. But when Dad lined the driveway with dahlias the size of dinner plates, our neighbors sniffed. Geraniums would have been okay, but dahlias were not. My eldest sister Marie's schoolmates taunted her for wearing hand-me-downs and wouldn't let her join their private after-school "club." When a classmate of my artistic sister, Yolanda, invited her over for drawing lessons, the little girl's mother told Yolanda, "We don't let Italians in our home," and turned her away at the door. Whenever something hurtful like this happened, Dad would tell us, "Don't get angry. Get a good, strong education. The only thing they can't take away from you is what's in your brain." Young as I was, this struck me as sound advice.

Fortunately, there are so often people who can see a child's potential, and Mrs. Emma Bowden, who lived next door, supplied the hideout that I needed. Often when I passed her house, she'd tap on her window. "I've made tapioca pudding," she'd call. "Come in. Let's do a jigsaw puzzle." With Mrs. Bowden at my side, the injustices of life that I was trying so hard to understand began to resemble a giant jigsaw puzzle. As I grew older, I started to see science in this way, too: nature would provide the pieces, and the scientist would figure out how they fit together to create a meaningful picture. And if someone else could solve a puzzle, then I was sure I could; if I was tenacious enough, I could even make sense of pieces that had seemed scattered and disconnected. Little did I know that this attitude—this refusal to give up—would not only help me when I *became* a scientist but would also help me *become* one.

Besides Mrs. Bowden, I had Miss Amy Striley on my side. Until I was twelve, I attended a four-room schoolhouse; Miss Striley was its principal. We were always taking exams, and in sixth grade I took what must have been an IQ test. Soon after, Miss Striley called me to her office. I was terrified. You went to the principal's office only when you had done something that could get you expelled. Carefully closing the door to her office, Miss Striley shook her finger at me. "Rita Rossi," she said, "you have a responsibility. You earned the highest score on this test. You *have* to go to college." I was so scared, I would have promised her anything. "Yes, yes, I will," I said—whatever it took to get out of that office. But Miss Striley didn't end her pitch there. My father took her evening English as a Foreign Language class, and she gave him the same message. Miss Striley's words became a bank of support that I would often draw from in the years ahead.

The children in our town had enormous freedom during the

summer. We'd finish our chores, pack a lunch, and leave the house to play. Only when it started to get dark would we dash back for supper. No one fenced off the beaches in front of their houses as they do today, so our dog Nippy and I could take long walks along the inlets of Beverly Cove. I also read voraciously in the town's excellent public library. During the school year, Miss Margaret Murray kept a stack of books in her fourth-grade classroom, and for every four you read, you could get one to keep. I loved words, so the first book I asked for was *Roget's Thesaurus*.

Freedom to wander during the summer gave me the freedom to pick my own friends. I liked anyone with a sense of humor, a sharp wit, a creative mind, and a willingness to discuss interesting things. I wasn't terribly fond of people who spent their time worrying about clothes or appearance. Without realizing it, I wound up with good friends from both sides of Beverly's railroad tracks. June's mother had a job as a nurse, but June told me her father was wacky because he went to church every Sunday but was drunk most of the rest of the time. When I visited June, it was clear there wasn't always enough food in the house. June and I caught frogs in a nearby brook or, when we'd saved enough money, went to Saturday movie matinees. My other best friend, Jean, had been born "out of wedlock," as they used to say, and her mother worked hard on the production line in the Sylvania light fixture factory. Jean and I read books in the library and spent hours talking about life and classical music. She would later marry a cellist in the New York Philharmonic orchestra.

I was fifteen when my life changed forever. The evening of March 29, 1950, my mother—my huggable fifty-one-year-old mother who sang Italian songs as she ironed and showed off my report cards to her friends at the bus stop—developed chest pains. My father and I took her to our family doctor, Dr. Leonard F. Box, who told her to go home and rest. The standard treatment for a man having a heart attack in the 1950s was complete bedrest in a hospital. Women weren't supposed to have heart attacks.

The next morning, I went off to school as usual. When I came back, my mother had finished the laundry and made lunch, and was sitting up, waiting for me to get home. We talked awhile, until, suddenly, her pain was so terrible that she went to lie down. I phoned Dr. Box, who said to give her paregoric, an opiate. I jumped on my bicycle, raced the mile to the drugstore, got the paregoric, and rode back like the wind. While I was gone, my mother tried desperately to call my father and brothers. Decades before cell phones, she could reach only one of my older brothers at his after-school job. He raced home in time to be with her when she died. I was too late.

Today I believe Dr. Box may have thought my mother was simply suffering from emphysema from gluing shoes in an ill-ventilated factory. In 1950, even if he had realized she was having a heart attack, there may not have been anything that could have saved her. Still, I had called Dr. Box about three o'clock, and we waited for him until he showed up around six. Then all he did was pronounce her dead. We also had to wait for Father MacNamara, the parish priest. When he arrived, I was sitting alone, shattered with grief. "Get up," he said. "Get over it." My father was sad and silent. And we children had no one to talk to, no one outside the family who could give us any support. Traumatized, my brother who'd been with our mother when she died took to his bed for days. *That's it*, I said to myself. *That's the last time I consider myself a Catholic*. I vowed to become a research scientist or a medical doctor to give poor and powerless people the care my mother was denied.

Returning to high school, I decided I could either show my anger or swallow it. Many of my friends had been giving themselves nicknames, so, shedding the name Rita (which I'd always hated), I chose a new one, Ricki, and a happy-go-lucky demeanor. Playing on the girls' basketball varsity team also defused some of my anger. I was five foot four, but scrappy. Three years later, our high school yearbook called me "the best good sport," declaring that, "When there is work to do, Rita is ready." The most accurate thing about me in that yearbook was that I wanted to go to college and become a "College Research Chemist."

The US Army and US Coast Guard helped educate my two older brothers as engineers, but Marie was designated the homebody who'd care for the family. She'd wanted to be a nurse, but my mother had said no, she should be a secretary, because secretaries didn't stand on their feet all day. Years later, Marie would go to night school and earn her bachelor's degree. Next in line was wry and funny Yolanda, who is six years older than I am and has had my back ever since I was a toddler.

Yolanda wanted to be an artist. That was fine. My parents revered Raphael and Michelangelo. Then my mother's friends warned her that artists studied nudes. So Yolanda had to be an art teacher, although she kept doing her own art, too. She proved to be extremely talented, showing her prints and paintings in well-established art galleries around the world. Then I came along, with my promise to Miss Striley.

After my mother died, our busybody aunt Brigida arrived each week—unasked and unwanted—to help with laundry and complain to my father about my wanting to go to college. Young women stayed home or attended secretarial school, I overheard her ranting one day, and after she left, I approached my father anxiously and told him, "I *really* want to go to college."

"Certainly you're going to college," he replied. "Look, I've never listened to her before. Why should I start listening to her now?" Years later, when I wrote my first book, he displayed it on the coffee table in his living room.

Studying college application forms my senior year, I gathered that would-be scientists needed excellent letters of recommendation from their science teachers. But this was the era of "No, girls can't do that." In high school, my brothers could play baseball and football, learn to fix cars, take shop class, and make electric lamps out of driftwood. I had to learn typing and cooking. My biology teacher made it clear that he preferred coaching football to teaching science to girls. Our school's physics teacher hated having girls in his class and, to my knowledge, taught only one his entire career—and it wasn't me. My chemistry teacher refused to write letters of recommendation for me and, I learned later, for some of my girlfriends. "Girls don't do chemistry," he told me matter-of-factly, a message I took personally, although he may just have been stating what was true at the time. Of the roughly four thousand chemistry faculty in the United States even twenty years later, only forty were female, about 1 percent.

The anti-female sentiment of Beverly High School's science program was not unusual. Astronomer Nancy Roman, known as the "Mother of Hubble" for her work on the space telescope, recalled asking her high school guidance teacher for permission to take a second year of algebra instead of a fifth year of Latin: "She looked down her nose at me and sneered, 'What lady would take mathematics instead of Latin?'" It's no wonder that 97 to 99 percent of the era's top high school graduates who did *not* go to college were girls. I wasn't sophisticated enough to recognize the prejudice in this—or the waste of human talent. My reaction was simply to figure out a way around the problem. I ended up asking a woman, my English teacher, for help. With her letter of recommendation, I applied to New England colleges that admitted women.

Of them, Smith College offered me no financial support and Radcliffe offered only $800 toward its $1,200 tuition. If I'd gone to Radcliffe, I would have had to live at home, work part time, and commute by train several hours a day to and from Cambridge. Also, although Radcliffe was Harvard's "sister college," women, including Radcliffe students, were banned from Lamont, Harvard's undergraduate library.

By this time, my family was rising in the world. My father had advanced from day laborer to foreman to founder of his own construction company, with wealthy and politically prominent clients like Senator Henry Cabot Lodge Jr. (whom my father privately called "Henry Cabbage Lodge"). Most crucially for me, my older sister Yolanda had married a physicist who was on a Fulbright Fellowship at Purdue University in Indiana, where she was teaching art. Yolanda knew something my high school should have but didn't: Purdue was the nation's biggest undergraduate engineering school, and its president was eager to attract top science students. My sister suggested I apply, and when Purdue offered me a full scholarship with room, board, books, and a way out of Beverly, I accepted immediately. My Harvard-educated history teacher was incredulous: turn down Ivy League Radcliffe for a public, Midwestern engineering school? But I've never regretted the decision.

• •

I'd never been outside the Boston area, but when I got off the train in West Lafayette, Indiana, in the autumn of 1952, I found myself in a massive construction site that reminded me of my father's work back home. The federal government was transforming public universities like Purdue into sophisticated research centers. World War II had been won with the help of scientific discoveries from Europe—atomic physics, radar, electronics, and computers—and Congress and the military did not want us depending on foreign know-how ever again. Purdue was awarded $48 million (about $500 million today) in construction money, while the G.I. Bill, which made it possible for many veterans to attend college, doubled the university's student enrollment to nine thousand. Some years there were nine or more male students for every female student.

Chemistry was my chosen major, but I quickly learned that lectures in that department were mostly about agriculture. And with 350 students per lecture hall, unless you showed up early, you ended up sitting so far back that you needed good binoculars to see the professor and the blackboard. Smaller recitation sections had perhaps fifteen students, but many of the instructors were German-born graduate students with heavy accents I could barely understand. They also had a habit of trying to date me.

I was so discouraged that I considered abandoning my dreams of science and medicine and switching my major to English literature. I took as many creative writing, poetry, and playwriting electives as possible—courses that have since helped me write eight hundred or more science publications and edit my students' work. I volunteered for student government and worked hard to turn around Purdue's less-than-stellar debate team, where I learned that the key to winning an argument is to assemble facts and more facts and then organize them in a rational way. I wasn't putting that lesson into practice in my own life, though. When a philosophy professor gave me a B on a paper and an A to a star quarterback who barely showed up for class, I walked into the professor's office, explained cogently that I deserved a better grade, tossed my notebook into his wastebasket, and left. He didn't change my grade, of course, and eventually I learned that uncontrolled anger makes your opponents resist all the more . . . but it's still a struggle.

Above all, I was annoyed when my ideas about science were not taken as seriously as those of the young men around me. In 1953, biologists had discovered that DNA carries the genetic code of living things. One day I asked my fungal genetics professor, "Why not use the DNA of microorganisms like bacteria and fungi to determine species?" Biologists do that routinely today, but my professor responded as if my idea were absurd. I wondered if that was why I was rarely called on in class, even when I raised my hand to ask a question.

I needed advice from other women interested in science. But now that World War II's manpower shortages had ended, almost no female faculty members remained in any science departments, and most of those who did felt too insecure to protest. Science funding was expanding job opportunities tremendously for men, but female assistant, associate, or full professors were rare. The 1960s were, according to historian Margaret Walsh Rossiter, golden years of government support for men, but the dark ages for women in science.

Most women working in American research laboratories had only master's degrees and functioned as handmaidens to male professors. Sexual predation was not uncommon, although we didn't yet have a term for it. When I learned that a prominent professor maintained a ménage à trois with his wife and an attractive undergraduate—and was seducing a foreign-born postdoc on the side—I wanted the dean of the college to know. Fellow students warned me to keep quiet. "Everyone knows," they told me. "No one in charge is going to do anything about him." Nor were women prepared to go public, much less band together to complain. Late one night at the prestigious Marine Biological Laboratory in Woods Hole, Massachusetts, a male graduate student grabbed a female colleague, tore off her shirt, and threw her to the floor. Slithering out from under him, she managed to escape. But it never occurred to her to tell other graduate students, her advisor, or the authorities. When Laura L. Mays Hoopes, who became a biology professor at Pomona College in California, finally spoke about this assault years later, she said that in that era, even if you told other women, "It was greeted with a wink and a nod."

Partly I felt alone at Purdue because in many respects I was. Almost all the undergraduate women at the university were majoring in home economics or nutrition. Biology, botany, genetics, and bacteriology had not yet been unified into the "life sciences," so we scientists were geographically separated from one another: zoologists working in one building, botanists in another, and none of them knowing the bacteriologists in the basement of a third. It also wasn't clear to me that a woman could have a viable career in science. My sister Yolanda had married a physicist, and he and his friends tried to convince me it was possible. They didn't have much evidence.

• •

Four women scientists were already laying the foundations of late-twentieth-century science. They were leaders in the genetic study of diseases, the structure of the atomic nucleus, DNA, and what came to be known as "jumping genes." But two of the four could probably not have supported themselves financially on their own.

When Czech-born Gerty Radnitz Cori and her Austrian-born husband, Carl, immigrated to the United States, they were told that working together was un-American and would ruin Carl's career. The Coris ignored the warning, and when I was in high school, they shared the 1947 Nobel Prize in biochemistry for showing how cells convert nutrients into energy. Working alone after her husband became a science administrator, Gerty trained six more Nobel Prize winners in her laboratory at Washington University School of Medicine in St. Louis and began the genetic study of inherited disorders. Until her Nobel Prize, however, she worked as a research assistant earning the equivalent of one-fifth of her husband's salary. Gerty worked in her lab until a few weeks before she died in 1957 of a bone marrow disorder, probably caused by the X-rays she'd used in experiments early in her career.

Maria Goeppert Mayer, who grew up in Germany, was responsible for our modern understanding of the atomic nucleus. Mayer was a beautiful flirt who loved to party and enchanted many of the men around her because she was so much brighter than they were. She fell in love with an American chemist—some said it was because he had the only convertible in town—and moved with him to the United States, expecting to have a career in physics. For the next three decades, she worked at three leading American universities as an unpaid volunteer, eventually working her way up to "voluntary professor." When she won the Nobel Prize in physics in 1963, her local newspaper headlined the news: "La Jolla Mother Wins Nobel Prize." (Lest anyone think no one would write about a woman scientist that way today, read the 2013 *New York Times* obituary for Yvonne Brill, the rocket scientist whose propulsion system kept communications satellites in orbit. "She made a mean beef stroganoff," it begins. To you noncooks, beef stroganoff is one of the easiest dishes to serve to company, and many working women bought cans of it in the 1970s.)

Two other women scientists—Barbara McClintock and Rosalind Franklin—were more employable because they were unmarried, but both ran afoul of the same man: James Watson. Franklin was on the verge of discovering the structure of DNA and the molecular basis of heredity by herself when Watson was shown Franklin's spectacular X-ray photograph of DNA's coiled structure without her knowledge or permission. He described the photo to his lab partner, Francis Crick, who had a background in analyzing crystalline structures. Reminded of the horse hemoglobin he'd studied for his PhD thesis, Crick realized that DNA's two coiled strands go in opposite directions: DNA was a *double* helix. It wasn't until 1999 that Watson publicly admitted, "The Franklin photograph was the key" to their discovery. Franklin died of ovarian cancer at the age of thirty-seven. The Nobel Prize is not given posthumously, so it went to Watson, Crick, and another DNA expert, Maurice Wilkins, four years later. Watson's subsequent bestseller, *The Double Helix*, turned Franklin—a strikingly good-looking woman with a sparkling wit and chic French tailoring—into an unattractive, inept spinster. A woman's appearance and age were important to Watson. When, at the age of thirty-nine, he married a Radcliffe sophomore, he sent a postcard to friends announcing, "19-year-old now mine." When asked in 2007 why it mattered how a woman looks, he answered, "Because it's important."

As for Barbara McClintock, after the University of Missouri told her she'd be fired if her mentor ever left, she stormed out and went to Cold Spring Harbor Laboratory on Long Island. When Watson later became director there, friends told her he'd called her "just an old bag who'd been hanging around Cold Spring Harbor for years." (Watson would be stripped of his job as the lab's chancellor in 2007, after remarks he made about the intelligence of African Americans.) Carnegie Institution grants kept McClintock financially independent of Watson, so she could continue to develop her revolutionary discovery that the chromosome is a fluid, moving, changing, and intricately regulated system in which genes migrate from one chromosome to another. Not that this discovery was appreciated everywhere at the time. The Purdue professor who taught my tomato genetics class in graduate school told us, "I have to discuss the 'jumping gene' theory, but the woman who developed it is considered crazy." In 1983, at the age of eighty, McClintock won a Nobel Prize for the theory—by then firmly established as fact.

These four women were stars, but their careers could not convince me that a woman like me could make a living as a scientist. It seemed as though these women were exceptions that weren't supposed to exist. At that point, I was trying to decide whether to go to medical school or get a PhD in science. I loved science—but as a physician, I could be financially independent and help people at the same time. Then, a laboratory technician in Purdue's brucellosis laboratory, where I had a part-time job my senior year, told me about Alice Catherine Evans, a bacteriologist who not only supported herself but also saved people's lives.

Shortly before World War I, Evans discovered that drinking unpasteurized cow's milk and handling infected animals could give people a chronic, painful, and potentially fatal condition known as brucellosis, also known as undulant or Malta fever. Evans's 1917 report set off a storm of protest from physicians, veterinarians, dairy industry representatives, and other bacteriologists. Evans was female, worked in a government public health lab, and had no PhD. She was the daughter of Welsh immigrant farmers in rural Pennsylvania and, as she wrote, her "dreams of going to college were shattered by lack of means." But it was less her lack of traditional schooling than her gender that posed the biggest problem to skeptics. If Evans was right, one male researcher said, a man would already have made the discovery. Only after men confirmed her research many times over was it accepted by the medical community and the dairy industry. Over the decades, her pioneering science has saved untold numbers of lives, and her work is recognized today as one of the most important medical discoveries of the twentieth century. (In an ironic twist, Evans was unable to attend her inauguration as the first woman president of the Society of American Bacteriologists in 1928, as she was hospitalized with the very disease she had studied.)

Even as a student, I felt a kinship with Evans. Her fascination with bacteria, her perseverance, and her magnificent work in public health made her a role model for me long before I knew the term.

• •

Surprisingly, I found my bearings in a Delta Gamma sorority house. One of my roommates there, Marilyn Treacy Miller Fishman, is still my most beloved friend. She would eventually become an eminent ophthalmologist, treating children's congenital eye diseases and training surgeons around the world, but during my junior year, she (as a senior) was preparing for life after graduation and debating whether to go to medical school or to get a PhD in science. I confided that I was discouraged by rote and overcrowded science classes. The interest I'd had in bacteriology, the study of living organisms so small they cannot be seen with the naked eye, had been considerably dampened by my introductory class, which was taught by a pompous, boring professor whose experiments had probably been designed in the 1930s and used unchanged every year since. "Before you make a decision," Marilyn suggested, "why don't you take Professor Powelson's bacteriology class? She's an amazing teacher."

Associate professor Dorothy May Powelson was one of the highest-ranking women in science at Purdue—in the entire United States, in fact. By 1960, the top twenty major research universities in the US employed only twenty-nine female full professors in the sciences, roughly one or two per school. At Purdue, Powelson taught advanced laboratory courses.

In my mind's eye, I can still see her. Dorothy Powelson was a very tall, pretty woman with twinkling eyes, a really nice smile, and a gentle manner. She moved gracefully, if somewhat hesitantly, from one task to another in the lab. She was perhaps forty years old and a strong feminist with a bachelor's degree and Phi Beta Kappa key from the University of Georgia and a PhD from the University of Wisconsin's prestigious bacteriology department. Our class was small and informal, a novelty in those days, with maybe six to ten students, half of them female. Powelson assigned each of us a 1000x light microscope in excellent condition and introduced us to almost all the bacterial species known at the time, from common ones like *E. coli*, which lives in the human gut, to really weird bacteria that form stalks and buds and grow at absurdly high and low temperatures.

"Look under the microscope—what do you see?" she asked me. When I looked through the lens, the elegance of the microbial world, with all its intricate structures, appeared almost miraculously. Those little critters wriggling around under that microscope enchanted me. *What are they? What do they do?* There were so many puzzles to solve; I was hooked. I decided immediately to switch my major to bacteriology, and I graduated in 1956 with a bachelor's degree with distinction.

I don't remember getting any personal advice from Powelson or having any private conversations with her. I would later learn she played the accordion and loved "all sports," sketching, and gardening. In those days, professors were supposed to stand on pedestals far above their students. We didn't chitchat with them or ask them for advice. But for me, just knowing that she existed was enough. Powelson influenced more women to enter the field of microbiology than any other person I know. Which is how I found myself on that beautiful May day, being told by Professor Henry Koffler that I had no future in science.

I didn't really like or even respect Koffler. Graduate student friends, both men and women, had told me he'd switched their thesis topics so often, it would take them almost a decade to get their PhDs. Worse, they were saying Koffler was among the men trying to push Dorothy Powelson out of Purdue. Her scientific background was as distinguished as his and she taught more advanced classes, but he had been promoted to full professor and she had not. The rumor was that she didn't bring in enough grant money. She moved to the Stanford Research Institute (now SRI International) in California. In 2008, twenty years after her death, two Stanford University men wrote an article that began with one sentence—only one—crediting her work with having started theirs.

Still, I was sure Professor Koffler would understand that I needed financial support for graduate studies. Like my parents, he was an immigrant who'd come to the US by himself, in his case, as a teenager from Austria. Assimilating, he'd changed his first name from Heinrich to Henry. He knew my grades were almost straight As. I felt he would understand.

But he didn't. And before Jack and I could continue on our way, Koffler added a parting blow: "The only degree you're going to get is in the maternity ward of a hospital." He said it in a voice that meant my case was closed.

Beside me, I felt Jack freeze. He knew I wanted a career and was determined to help me succeed. Fuming—silently, so Koffler wouldn't think he'd crushed me—I made a promise to myself: *I will get my degree*. Adrenaline surging, my mind raced ahead, trying to figure out a strategy. I had been accepted to three medical schools but had been able to obtain deferred admittance so I could stay at Purdue and earn a master's degree while Jack finished his. I worried that Koffler would prevent this plan.

And so I turned to the one male professor I trusted: my undergraduate advisor, professor of genetics Alan Burdick. I told him what Koffler had said.

Now, Alan Burdick was a good and very insightful scientist, but he wasn't one of Purdue University's favored few. So he smiled and said five words I'll never forget: "Their loss is our gain." He went on to tell me he needed someone to manage his collection of *Drosophila* (a genus of small flies), and offered me a spot as a research assistant in his lab. So every week for the next year, I prepared a sweetly aromatic mix of molasses, yeast, and corn flour to feed to Burdick's stock of fruit flies. Visiting me in the lab, Jack discovered that fruit flies have mites that bite and made him itch for hours afterward; I had grown used to them. Genetics wasn't my first love—that was bacteriology—and yet ironically, the genetics thesis I'd write for Alan Burdick would prepare me beautifully for the twenty-first century. But I'm getting ahead of myself.

Koffler went on to become chancellor of the University of Massachusetts Amherst and a popular president of the University of Arizona. Years later, when asked about what he'd said to me, he denied any bias against women but didn't deny his comments. "I'm not about to deny it because I think Rita is definitely truthful," he said.

After we'd finished our master's theses, Jack and I moved blithely on to the University of Washington in Seattle, one of the few places that admitted us both for postgraduate studies. I had been accepted to the University of Washington School of Medicine, but I couldn't be admitted until I'd been a legal resident of the Pacific Northwest for a year. So I settled for starting a PhD in biology instead.

While I still had a lot to learn about being a woman in science, I already knew one of the first rules for being one: there were some real heroes out there. I just had to find them.

chapter two

Alone: A Patchwork Education

Each incoming graduate student at the University of Washington was assigned a faculty member who was supposed to function much like a guidance counselor. Mine was a microbiologist with a PhD in winemaking—not quite the research subject I had in mind.

Nevertheless, I showed up at his office in September 1958, eagerly expecting him to steer me toward a PhD mentor who would guide my research dissertation and my future career. Introducing myself, I helpfully explained that one of my Purdue professors had suggested I work for geneticist Herschel Roman. The advisor's face fell, and he iced up immediately. I was orphaned before I could even start my PhD.

Outside his office, I asked some longtime graduate students what I'd done wrong.

"You've made a terrible mistake," they told me. I'd gotten myself stuck between two feuding professors. They mended bridges later in their careers, but at that point, they hated one another, I learned.

Fortunately, Alan Burdick had kindly already written to Herschel Roman on my behalf, so I went to Roman's office and he agreed that I could work in his lab. I stuck it out for two semesters, during which he helped his male graduate students with their research and theses but gave his female technician and me brusque orders with no chance to ask questions or contribute intellectually. Then, because doing a thesis with Roman while remaining a graduate student in microbiology was proving bureaucratically difficult, I decided I had to leave his laboratory.

But without a mentor for my PhD, my career was over before it had even begun. I'd thought talent, hard work, and good research would be enough for me to succeed in science, but was finding that maybe it wasn't.

• •

I should explain that graduate school professors in science are all-powerful. They admit students to their laboratories, fund their research, pay them a stipend, and approve (or reject) their theses—all according to their own personal rules and expectations. With the advent of federal grants, professors accepted students whose careers would extend the luster of the professor's own work. Women, who most people assumed would get married and quit when they had children, weren't considered worth the time and money it took to train them. As a result, American universities maintained—openly, unapologetically, and legally—two separate tracks for students: one for men, one for women. The men got the top PhD degrees, great jobs, and nearly all the research money. The women got master's degrees that prepared them to work as technicians in scientific and medical laboratories run by men. A lucky woman could teach introductory college classes but not as a professor.

The system was casual, collegial, and closed. A few years before I began my PhD, the University of Washington found itself in need of a geneticist, so a professor wrote around to some plant geneticists he knew to ask if any "young man" was available. The future Nobel Prize winner Barbara McClintock, one of the all-time giants in genetics (whom you'll remember from the previous chapter), had recently left the University of Missouri in a fury after being told she'd be fired if she ever dared to marry. "Of course the number one person in the world in this field is Barbara McClintock," one geneticist told his department chair. "It's too bad you can't hire her because she's a woman." Instead of McClintock, the University of Missouri recommended Herschel Roman, the man who became my tormentor, and the University of Washington hired him sight unseen.

Discrimination against women was nothing new, of course. But its scale in the 1950s and '60s was unprecedented, because, encouraged by the federal government, the number of American women graduating from college was doubling. The biggest social change in world history—the de-gendering of occupations—was under way. Still, there were too few of us to spot discriminatory patterns—or, if we did identify them, to fight them.

Again, graduate students clued me in. Four or five other women who'd earned master's degrees while working as teaching and laboratory assistants in the microbiology department at the University of Washington had been pushed out of the PhD program in recent years. (All were able students who, after going elsewhere, succeeded in earning PhDs, medical degrees, or having otherwise distinguished careers.) According to Margaret A. Hall's history of women at the University of Washington, one graduate student even threatened to pursue legal action against Herschel Roman, accusing him of unjustly edging her out of a PhD program. Taking legal recourse was unthinkable to me. Still, I came to realize that no one in microbiology or genetics would want a woman graduate student—especially not one who had walked out of her previous lab. And because I still had not yet lived in the Northwest for a full year, I couldn't enroll in the University of Washington's medical school, either, even though I could—and did—teach the university's medical students genetics and bacteriology as a teaching assistant.

If I'd been able to read Hall's thesis (which was not completed until 1984), I would have understood more about what was happening in Seattle. In theory, Washington was a good place to be a woman in science. Twenty percent of US universities—top places like Princeton and Georgia Institute of Technology—awarded doctorates only to men. But in her history, Hall documents how during the first half of the twentieth century, the University of Washington's administration consciously masculinized its faculty in an attempt to "upgrade"—it was 85 percent male by the time I arrived. Women scientists were segregated into "female" fields like home economics, nursing, and women's physical education, while women in other departments were rarely permitted to advance above the level of low-paid "instructors." Hall and her husband, Benjamin D. Hall, an eminent geneticist at the university, believed her meticulously documented history got her blackballed. She was never hired as a faculty member at any university, and her husband would say her thesis, written at the peak of the women's movement, strained his relationship with some of his own colleagues.

What confused me was that four women scientists had managed to make names for themselves at the University of Washington. Two of them did it by building support from outsiders, thanks to a new and unfortunately short-lived technology: educational TV. Erna Gunther, chair of the anthropology department, popularized the art of Northwest Native Americans around the world, hosted a regular radio and TV series called *Museum Chats*, and built the Washington State Museum of Anthropology. Her fan base was so loyal that the university's attempt to oust her from her museum directorship caused a statewide scandal. Marine scientist Dixy Lee Ray hosted another public TV program, *Animals of the Seashore*, and saved a nearby river delta as a wildlife refuge. She won a prestigious Guggenheim Fellowship and raised millions for marine science from the National Science Foundation, and yet the university's zoology faculty twice voted to deny her tenure, claiming she hadn't published enough. When Ray was elected governor of the state of Washington in 1976, she thumbed her nose at the university's budget requests.

Two other women scientists who might have mentored me occupied such lowly positions on campus that I felt I could hardly consider them authorities on academic career building. Dora Priaulx Henry was a world expert on barnacles, and Helen Riaboff Whiteley was the university's star microbiologist. Yet both were "associates" because their husbands were professors at the university, and the state's anti-nepotism laws and university regulations forbade hiring relatives. Modern anti-nepotism rules say simply that relatives cannot supervise relatives. However, for most of the twentieth century, universities and colleges in the United States enforced anti-nepotism regulations almost exclusively against wives of faculty members; exceptions were routinely made for brothers, sons, and nephews. The system was particularly difficult for women scientists, because then, as now, so many of us married other scientists, no doubt because of shared interests and time spent together in research programs. The state of Washington's anti-nepotism law was especially draconian, as it prohibited wives from paid work anywhere in the university except as clerks, secretaries, or laboratory assistants. Several women faculty members were dismissed after they married colleagues.

Helen Whiteley's low-ranking position was particularly ludicrous because the university's entire chain of command confirmed her status every single year. First, the chairman of the microbiology department signed paperwork to renew her contract as a research associate. He then sent the paperwork to the medical school dean, who signed it and forwarded it to the university president, who likewise signed it and forwarded it to the board of regents for approval. And what's worse, Whiteley and her husband were proud of this help. Fortunately, in 1965, shortly after I had earned my PhD from the University of Washington, life changed for Whiteley. The National Institutes of Health (NIH) awarded her a Research Career Development Award to pay half her salary for the rest of her career—but only if she were a tenured professor. Eager to get its hands on the money, the university forgot its own rules and immediately jumped Whiteley four rungs up, from research associate to assistant professor, associate professor, and finally full professor. After Whiteley cloned a gene that made cotton and tobacco crops naturally insect-resistant, she had every reason to hope she'd be elected to the National Academy of Sciences. When she wasn't, it was rumored that a member of the academy had blackballed her.

Could I have gone to Helen Whiteley for mentorship during my time in Seattle? Well, no. Even her adoring husband, Arthur, described her as "stern," and said his wife disapproved of the women's movement because she hadn't had help and believed that any woman who was really good at science didn't need any. Whiteley took no female graduate students until the last years of her life, when she had two.

In truth, both Helen and Arthur Whiteley were stern. For years, the Whiteleys, who had no children, had dinner at their house every Friday with three "bachelor" faculty members: two men, and Dixy Lee Ray. Some outsiders called the group a clique. But when Ray became an outspoken proponent of nuclear power, Arthur told Ray she was no longer welcome in their home. Helen Whiteley was not someone I could ask for help.

• •

You couldn't predict whether being married would help or hurt your career, either. Frieda B. Taub and I arrived at the University of Washington within a year of each other, but she came with a doctorate. Up until Seattle, her husband had been an asset to her training and career.

Many academics feared that having a woman in their lab would cause a scandal; female secretaries and technicians were okay, but not scientists. Taub's PhD advisor at Rutgers University in New Jersey told her he had taken her on as a student only because, since she was married, his wife wouldn't object. When Taub had to attend a weeklong field trip for a course, another professor promised her husband she'd get an A if he accompanied and chaperoned his wife. Still another professor refused to co-chair a committee with Taub because his wife objected to their traveling together.

In Seattle, though, her marriage was a clear liability. Several department chairs at the University of Washington refused to hire her because she was married, even though her husband wasn't faculty (he was a systems analyst at Boeing). And even after the university hired her, the prejudice against professional women didn't stop. Taub became the first working wife allowed to adopt a child in Seattle; before, would-be adoptive mothers had had to promise not to get a job until the child's eighteenth birthday. The discrimination was all "quite open," she said later. "Any woman at the time knew she was in a man's world and it was going to be tough, so there was no point talking about it. You knew what you were getting into."

As a result, we worked alone. We didn't have little groups of women working together, sharing problems and successes, taking risks, and believing in ourselves. We knew we needed support—but we needed the support of men. John Liston, a wonderfully caring man who would become my PhD advisor, rescued two other women scientists, Taub and Joyce C. Lewin, by finding jobs for them in the oceanography and fisheries departments when no one else would hire them; Lewin specialized in diatoms, and Taub studied the ecology of aquatic communities. Taub would eventually become a full professor. When she gave a talk at the university in 2019, more than fifty years later, the male professor who introduced her led by mentioning the date of her marriage and making nice comments about her husband and children. Fortunately, some things do change: the university had a woman president at the time, and the School of Aquatic and Fishery Sciences, which videotaped Taub's talk, deleted the introduction because the faculty deemed it inappropriate.

• •

In my early years at the University of Washington, I began to learn of some unsettling events that had occurred at Purdue after I left—events that told me the fight for equality in science would be long and difficult. Three of these stories emerged only in bits and pieces over the decades. I followed each closely because they'd occurred at my alma mater—and because if I'd stayed there, something similar might have happened to me.

The first story pitted Holocaust survivor Anna Whitehouse Berkovitz against Henry Koffler, who, three years after denying me a fellowship, became chair of the Department of Biological Sciences, which united all the life sciences under his purview.

Anna Berkovitz was thirteen years old when the Germans arrested her family and sent them to Auschwitz. From there she was sent to Birkenau and then to a slave labor camp near Magdeburg, Germany, where she worked in an underground ammunition factory until the Swedish Red Cross freed her. Only Anna, one of her sisters, and her mother survived the war. Forever after, Berkovitz felt she had to justify her survival; she could not face leading a useless life. When her husband got the job of his dreams at Purdue in 1962, she started working toward a PhD, taking classes at the university while her two sons were in school. Later, her husband took a sabbatical year in England, and she enrolled in the world's most advanced human genetics program at the Galton Laboratory, University College London, where she studied from nine to five, five days a week.

Upon returning to Purdue, she resumed work on her PhD, only to learn that her world-class genetics training had changed Koffler's estimation of her worth. Calling her into his office, he said, "You can't get a PhD; you have to teach the genetics you've learned in England." The department expected her to teach others to do the research she wanted to do herself. "But can't I teach and finish my PhD at the same time?" Berkovitz asked. That seemed simple enough. As a refugee au pair after the war, in the space of two years, she'd acquired proficiency in English, high school and college diplomas, and a Phi Beta Kappa key. "Lots of graduate students teach while they earn their PhDs," she told Koffler. "Why not me?"

Koffler said no—if she got a PhD, she'd be unemployable. As the wife of a faculty member, she'd never get a job at the university—and without a recommendation from Koffler, she'd never get a job anywhere else, either. Her choice: get a PhD and never get a job, or teach without a PhD at the bottom of the academic ladder. Anna's husband was also Jewish, and in 1965, the likelihood of his getting a good position elsewhere was small. Besides, she was told, "Your husband is a full professor in another department. You don't have to worry about money."

The choice wasn't really a choice, of course, and she abandoned her pursuit of a PhD. For the next thirty-five years, Anna Berkovitz ran genetics lab courses, often teaching 450 to 500 students in 10 to 15 sections. She designed new courses, had the heaviest teaching load in her department, and earned sixteen "best teacher" awards—but was never once promoted. Instead, she was given tenure—a lifetime contract—in her lowly position as an instructor.

When Berkovitz retired in 2003 at the age of seventy-three, the department faculty gave her a banquet and a certificate designating her an instructor emeritus. In her speech at the banquet, she spoke of how frustrating it was to do a good job but never get promoted; her colleagues subsequently gave her another certificate, this one naming her a professor emeritus. In her entire career, her salary had never approached that of a newly hired male or female assistant professor. After her retirement, she said, the university hired four people to replace her.

The second disturbing story from Purdue concerned Violet Bushwick Haas. Haas's husband was hired in 1962 to build a first-class math department at Purdue. He was given free rein to hire twenty-one mathematicians, anyone he wished—anyone, that is, except his equally stellar wife, who had a PhD in math from MIT. Fortunately, Purdue's anti-nepotism rules were sufficiently lax that although she could not work in the College of Science, where her husband was the dean, she could be given a position in the electrical engineering department. This seemed like a good solution, until she discovered that the all-male engineering department deeply resented her. She was assigned a small closet as her office space, and if the entire engineering department was writing a grant proposal together, Haas and only Haas was excluded. Only when Haas threatened to resign was she promoted to professor.

Haas fought hard for women at Purdue. She formed a group of women in science and engineering that met monthly to discuss problems and possible solutions. Anna Berkovitz attended regularly, and Haas was the woman Berkovitz went to whenever she had a problem. In 1983, when Haas spent a year at MIT with a National Science Foundation program for visiting women professors, she raised hell when she discovered an annual male student "tradition" that involved showing an X-rated film during registration week. Her objections helped end the tradition.

West Lafayette, Indiana, was a one-company town in the 1960s. Before the federal highway system and cheap commuter airfares, Anna Berkovitz and Violet Haas were trapped in whatever jobs Purdue offered. They couldn't leave to seek employment elsewhere without causing turmoil in their marriages and family lives.

But what if a man and a woman wanted it all: marriage, children, and professorships in the same field? A third story trickling out of Purdue involved my very good friend and former classmate, J. Alfred "Al" Chiscon, and his wife, Martha O. Chiscon. They were both biologists and extraordinary teachers, recognized nationally for their classroom innovation; as a pair, however, they were unemployable anywhere other than Purdue. When the University of Texas phoned Al one day and said, "We'd like to make you an offer," Al asked, "Which one of us?" Texas answered, "Well, either one, but we only want one of you to get paid."

The Chiscons had had to get Purdue's permission to marry. In 1969, they went to see Henry Koffler and asked, "If we get married, will one of us be required to leave?" Martha was teaching one of the nation's first courses on women in science, and her students were telling heartbreaking stories. To stay in science, some had had to forgo marrying the man they had wanted to marry. Others lived with their partners but without marrying or having children, and still others remained single. But times were changing. The women's movement was spreading to science, and more women were demanding the right to pursue careers. I think Koffler was politically astute enough to know he had to give in. So the Chiscons married, had three children, became professors of biology, and taught more than sixty-five thousand Purdue students during their careers. Martha also became an associate dean of science—but earning $35,000 less than her male counterpart. When she stepped down, she was replaced by three people, each of whom earned more than she had.

• •

As for me, I could see no way forward at the University of Washington except to do what I'd almost done at Purdue: abandon science and earn a degree in English literature. I planned to study sixteenth- and seventeenth-century English poetry—the metaphysical poets—inspired by Andrew Marvell's poem "The Garden," which reads like a hymn to ecology:

The mind, that ocean where each kind,
Does straight its own resemblance find,
Yet it creates, transcending these,
Far other worlds, and other seas;
Annihilating all that's made,
To a green thought in a green shade.

I had wanted to explore the modern science behind that poem: to understand the garden of the senses, the mind and soul in nature and human life, and how to expect more than the narrow verged shade. But orphaned as I was, it seemed that I'd never get the chance.

I was frustrated beyond belief, and not very good company for Jack. But soon I heard that a young man from Scotland had just joined the University of Washington's Department of Fisheries and was seeking a technician to set up his laboratory. I needed the money, so I took the job. I selected and ordered the equipment he'd need, oversaw the carpentry, and helped him hire a technician. After watching my father plan construction projects, none of this was difficult. In a few months, the Scotsman, John Liston, was ready to start work.

As it turned out, Liston approved of self-starters. He even chuckled when I politely—and rather kindly, he thought—set him straight about the work I was doing. One day, he informed me I shouldn't be his assistant. I should be his graduate student.

Liston told me that marine bacteriology was just getting started, so the field was wide open, with very little competition. He himself was one of perhaps a half dozen marine bacteriologists in the world. The university's fisheries department wanted him to help the state's salmon industry by studying fish diseases and causes of spoilage. But Liston, who'd had a rigorous education in biochemistry in Scotland, had already negotiated permission to offer a PhD in marine microbiology. I would be the first student in the program, and he warned me that, based on his wife's experiences with Scottish fishermen, both oceanography and fisheries attracted "hunter" types. The University of Washington was one of the few oceanography departments in the US to allow women on research cruises, and Liston promised to get me onto research vessels, if only for the day, as the opportunity arose. (Rumor had it that an early department chief had changed the rule regarding women on board because he'd wanted to bring his mistress.)

I was naive. I barely knew the value of having a mentor—or even what a mentor was. I only knew that some women scientists were alone and powerless, and some were ridiculed. I wanted a PhD advisor who'd treat me with respect and let me study the genetics of bacteria. And for someone like me, who'd grown up on the Atlantic coast, *marine* bacteria sounded fascinating. I'd be studying microbes in their natural environment, uncovering their life cycles and how different species fit together in the web of nature. I quickly agreed to become Liston's first PhD student and changed my focus for the fifth and—almost—last time. I'd studied chemistry, English literature, bacteriology, medicine, and genetics. Now I'd be investigating oceanography and bacteria associated with marine animals, including fish, shellfish, and invertebrates. And in the months to come, when the fisheries students—all men—joked about my standing on a box to reach the sink when I dissected fish to study the microorganisms in their guts, I gave as good as I got. Then they called me a "smart cookie."

Liston turned out to be a wonderful mentor: enthusiastic, unconventional, and rebellious. The Scotsman was only ten years older than most of his students—so, as he put it, "we could interact like normal human beings." At parties, he could down a fifth of Scotch and stand to sing "The Northern Lights of Old Aberdeen." He was an avid cricket player and an excellent scientist, whose attitude was that every rule was meant to be broken.

Early one Sunday morning during my second year of graduate work, Liston phoned me and said, "I'm ill. I'm scheduled to give a talk in Philadelphia this week at the American Society for Microbiology's annual meeting, and I'm too sick to fly tonight. You're going to have to give the paper. You're coauthor on it, anyway. You can read my notes on the plane." So I did. I've since realized he probably wasn't sick. I think he was fixing things so I could add presenting a paper at a scientific meeting to my résumé.

John Liston had no prejudice against women; he was a strong supporter of women in science and delighted in telling me about brilliant women scientists like Emmy Klieneberger-Nobel, who had been considered "dotty" by her (male) peers. She would arrive at meetings of the Royal Society in London with a briefcase full of slides about her pioneering work on mycoplasmas, bacteria without cell walls. Speaking with a strong German accent she called her "Continental voice," she'd proceed to show the stack of slides. Klieneberger-Nobel may have been considered "dotty," but she certainly was not. She was a Jewish refugee from Nazism and later wrote in her memoir, "If my family had not perished so tragically under the Nazis I could be completely happy to the end of my days in England but, of course, this cannot and should not be forgotten and, even if subconsciously, it is always there in the background of my mind." In addition to being a victim of Nazism, Klieneberger-Nobel was a victim of sexism. Liston would point to the tragedy of a woman scientist trying to be recognized for her insights and being treated badly instead.

• •

When I began my PhD work, a major goal of microbiology was identifying bacteria to make sense of what we were calling species. Traditional taxonomists, especially those trained in the intricacies of Latin and Greek scientific nomenclature, would quarrel over the names of species and cluster organisms according to their appearance under the microscope. This was difficult work, because under the light microscope used for early microbiology, many bacteria looked alike. For my thesis, I included an extensive array of biochemical and physiological tests to analyze a bacterium's ability to ferment sugars, break down proteins, and grow and metabolize at extreme temperatures. I applied these tests to so many bacterial species and strains that I wound up with enormous amounts of data about microorganisms no one had published about before. After learning of an English scientist who was using a computer to find similarities between organisms based on shared morphological and biochemical characteristics and then classifying plants and animals according to those results, I decided to try the University of Washington's first "high-speed" computer, an IBM 650. This wonder of a computer—the size of three large refrigerators with the storage capacity of a microwave oven's microprocessor—was housed in a gazebo atop the university's chemistry building. Graduate students were allowed to use the 650 only between midnight and six a.m. The university did not yet teach computer programming, but a Canadian postdoc lab mate of my husband's, George Constabaris, had used a computer when he worked in the chemical industry, and he kindly taught me programming. Using machine language, I wrote the first program to identify bacteria isolated from their environment. I had to punch a separate IBM card for each strain and, with no technician to help, wire the computer's board to run the program.

The prestigious journal *Nature* published my article about the work in 1961. I wasn't a computer genius, but I could see that computers weren't just tools for higher-order calculations; I knew they'd revolutionize the sciences. The University of Washington professors on the committee that would pass or fail my dissertation knew next to nothing about computers, so I included my software code in the thesis and glued two IBM punch cards to page 41. No one at the university thought to patent the software. We were working for the public good. The science I was doing was truly exciting.

One of the benefits of specializing in an esoteric topic like identifying marine bacteria was that—supposedly—I wouldn't be competing with giants in the field. For my thesis, I'd planned to focus on one particular bacterium, *Pseudomonas aeruginosa*, that I had isolated from samples I'd collected. The bacterium is commonly found in water and soil and known to be dangerously resistant to antibiotics. When Roger Stanier, one of the leading bacteriologists of the day and an expert in *Pseudomonas* bacteria, invited me to give a talk at Berkeley on the subject, I was looking forward to being among other people who would share my passion for marine bacteria. A few minutes into my talk, though, Stanier began speaking over me. At first I thought he had something he really wanted to say, so I politely waited for him to finish. Then he started criticizing my results. I continued with my talk but I was confused and angry. It took me years to understand why anyone would publicly harass a young scientist and ridicule her work, but I now know what set him off: I was not *his* student, not part of *his* laboratory, and I was intruding on *his* bacterium, about which he was clearly an expert. Would he have treated a male student the same way? Probably not. More likely, he would have been constructive in his criticism and actually helpful.

I couldn't ignore the possibility that I'd have to face Stanier again at scientific conferences—a chance criticism at a national meeting, and my career would be on the line. I decided to get out of his way by finding other microorganisms to study, changing my research focus for the sixth time. In retrospect, Stanier did me an enormous favor. Because of him, I switched to vibrios, some of the most common bacteria in the aquatic environment, especially surface coastal and ocean waters. It turned out that some vibrios are dangerously pathogenic to humans—and extraordinarily interesting.

Changing fields was supposed to be bad for a scientist's career, but it taught me how to apply interesting ideas from one field in another field—for example, to use techniques from yeast genetics and fruit flies to study the genetics and ecology of microbes in the marine world. It was a patchwork education, but it taught me to see how big, natural systems worked. As a result, I became a molecular microbial ecologist and a champion of holistic science and interdisciplinary interdepartmental research teams, something almost unheard of at the time. A project that started in the crucible of despair turned into one of the best decisions I ever made. Now, fifty years later, that decision has kept me active in one of the hottest fields in the life sciences: microbiomes, the study of all the genetic material in all the microorganisms in a particular environment, whether it be the human gut, food, river water, or the ocean.

In the end, my thesis on the bacteria that live in marine animals was approved, not by the microbiology department or the oceanography department, but by the School of Aquatic and Fishery Sciences. "It's splitting hairs," Liston said. I'd been in his marine microbiology PhD program, whatever the bureaucrats in the fisheries, oceanography, and medicine departments said.

• •

As Jack and I were completing our PhD studies, Jack was told that Canada's National Research Council in Ottawa would be an excellent place for a postdoctoral fellowship in chemical physics. We both applied for postdocs there and were delighted when we both received congratulatory letters. But sometime later, I got a second letter, and this one, from council chair Norman E. Gibbons, did not have good news. Anti-nepotism rules, Gibbons wrote, would prevent the council from awarding both Jack and me fellowships. He didn't have to say that the National Research Council assumed the husband should get the money; I knew that immediately. After all, Jack had received no such letter.

Once again, I was upset and disappointed. But with this setback, unlike those I'd experienced earlier in my career, I had no thought of leaving science. For some people, myself among them, science is the most exciting professional endeavor imaginable. Women knocked—indeed, pounded—on the doors of science because we loved working in the laboratory and in the field, making discoveries, learning new principles, and understanding how nature works. Shirley M. Tilghman, the first woman, and the first biologist, to become president of Princeton University, described the feeling well: "My first big discovery was literally electrifying," she wrote. "My heart beat wildly. The hair on the back of my head stood on end." After that, she said, "nothing could have stopped me from becoming a scientist." I felt the same way. But without a laboratory, I couldn't work as a scientist.

Soon, however, a *third* letter arrived from Ottawa. This time, Gibbons had good news. He and his colleagues at the National Research Council studied bacteria that require salt for growth and metabolism; the bacteria I studied mostly lived in salty water. Gibbons, a kind and generous man, said he'd be pleased to offer me free laboratory space and access to all the equipment and reagents in his stockroom.

When I showed Gibbons's letter to John Liston, he set to work.

(And looking back, I now realize that Liston must have talked to Gibbons on my behalf; the two were friends.) In a masterful bureaucratic sleight of hand, Liston applied for a National Science Foundation research grant for both of us, but generously named me, then age twenty-six, co-investigator. Next, he convinced our University of Washington dean to appoint me to a research assistant professorship and grant me a leave of absence to work in Ottawa. It was a splendid example of the administrative skullduggery (not to mention mentoring) needed to keep a woman in science in the 1960s. This was 1961; the structure of DNA had been elucidated only six years earlier, when I was an undergrad, and using computers to study the genetics of marine bacteria was a new field. For the next ten to fifteen years, I would be the only microbiologist funded continuously by the National Science Foundation (NSF) to study evolutionary relationships among marine bacteria. Jack and I were excited to begin our new lives in Canada and drove out of Seattle forgetting to pick up our diplomas.

In Ottawa, the NSF grant enabled me to hire a laboratory technician, someone to assist in carrying out experiments and preparing reagents. I hired Margaret Briggs Gochnauer. Gochnauer was sixteen years my senior, and her advanced degrees and experience were more prestigious than mine. Raised as an only child by a single mother who worked a department store job to support them, Gochnauer put herself through San Jose State University in California despite severe dyslexia and did summer research at Woods Hole in Massachusetts and the Hopkins Marine Station in California. She wanted to have it all—scientific research *and* a house full of children—and married her husband, Thomas, in 1950. For her master's thesis at Stanford University, she discovered and described the lifestyle of two species of hermaphroditic roundworms later named after her: *Rhabditis briggsae* and *Caenorhabditis briggsae*.

Gochnauer and her husband moved to the University of Wisconsin, where her husband was doing a PhD on bees. But the only advisor she could find for her own thesis was Elizabeth McCoy. This was a problem—McCoy believed that training married women was a waste of time because too many quit when they got pregnant. So Gochnauer kept her marriage secret and got her PhD.

By the time I met her, she was desperate for a job—as expected, once authorities learned that she had three children, she was deemed unemployable. Because my grant came from the United States and didn't trigger Canadian anti-nepotism rules, I could hire her. So there we were, two victims of North America's anti-nepotism rules, working happily side by side in my tiny lab as I consolidated in my mind the idea that vibrios—including cholera vibrios, which would later become the focus of my career—are marine creatures that live and thrive in water. After I left Ottawa, Gochnauer was never again able to secure a long-term research position. She was another talented, underappreciated woman scientist.

By my second year of the postdoc, I was pregnant with my first daughter, Alison. One day, microbiologist Donn Kushner, a very kind man with whom I shared the lab, said to me, "You're expecting. Do you have any help lined up?" I did not expect to have maternity leave—if you asked for maternity leave in those days, you'd be given *permanent* leave—but Jack and I hadn't thought about hiring help, either. Fortunately, Kushner was an old hand at childcare arrangements; he and his wife, a professor (and future university president), had three boys. He recommended his housekeeper's friend, Kit Godson, and one day this English lady arrived for an interview with a little notebook and questions like "When do you have breakfast? What are your plans for the child?" and other things you'd ask the lady of the manor. Godson's mother had died when she was sixteen, and when her father remarried, his new wife kicked Godson and her six siblings, one of whom was still a toddler, out of the house. Godson took care of the youngest while the others found work. When all her brothers and sisters were on their own, she became a professional nanny. She remained with us for seven years. It was Kit Godson who made it possible for both Jack and me to have careers and a normal family life. Our two daughters loved Miss Kit, and after she retired to England, we visited her almost every summer until her death.

• •

With Jack's Ottawa fellowship winding down, I attended the annual meeting of the American Society of Microbiology. Then, as now, the annual meeting was both a scientific conference and a job fair for young scientists, and I was hunting for a permanent faculty position. I ran into Dick Morita, a very good friend and colleague who was then at Oregon State University, and asked if he knew of any openings; he pointed to a small notice pinned to a bulletin board: "Newly hired chairman of biology dept at Georgetown is seeking essential faculty members in microbiology." The chairman was a friend of his.

I was interested. Georgetown is located in Washington, DC, where Jack had just accepted a physics position at the National Bureau of Standards (now the National Institute of Standards and Technology). I met George Chapman, Georgetown's biology chair, in the lobby of the hotel where the meeting was being held. He was under the gun; the university's president had ordered him to double the size of his department, establish graduate programs, start research, and, of course, obtain grant money—all within a year. I told him I had grant money, had published several papers, and had data for several more. He offered me a job on the spot. We shook hands, and that was that: no interviews or guest lectures. Morita was my reference.

I'd given myself ten years to finish my PhD and postdoctoral fellowship, and I'd done it in eight. I was set—I thought. Georgetown was Jesuit, and the school did not go coed until 1969; I was a lapsed Catholic and would be the only woman on the biology department's faculty. But neither Chapman nor Georgetown's president cared. Few of Chapman's hires were practicing Catholics, and he was accustomed to strong women; his widowed mother had worked menial jobs to get him his Princeton education. As a young faculty member at Harvard, he'd taken on a female graduate student, Sarah P. Gibbs. Her first advisor had asked her where she wanted to be in ten years. When she said, "Just where you are," he advised her to give up all dreams of a PhD and fired her. "Prepare a high school curriculum or a manual for a freshman biology course instead," he told her. The next day, Gibbs walked into Chapman's office and he agreed to supervise her PhD studies. Gibbs went on to become a professor at McGill University in Montreal.

Chapman was a kind and supportive boss. When I told him that Jack loved racing sailboats and wanted me to crew for him on weekends instead of working in the lab, his response was, "Go. Keep Jack happy." In 1965, Chapman came to the hospital while I was in labor with our second daughter, Stacie—although, truth be told, the dear man also wanted to collect the placenta for an experiment.

Georgetown's biology department wrapped me in a friendly cocoon for almost a decade. As long as I published research, I did not have to worry that a controversial scientific theory might affect my career. I was free to "find out" on my own how nature works. Had I been at Harvard or at Caltech, I would have been competing with male powerhouses and constantly dealing with intellectual jousting. At Georgetown, half the biology faculty dated from the years when the department had done little other than teach undergraduate courses. One faculty member from the "old guard" complained that my writing computer software to analyze species was "purely mechanical." He had no idea that computer programming involved mathematics and abstract reasoning.

With two young daughters and a husband busy with his own career, I couldn't take the year-long sabbaticals that used to be standard for academics. Instead, I traveled one or two weeks at a time to other laboratories, domestic and abroad, finding collaborators and bringing news of the latest developments back to my students. Because of the children, I never took extra time for sightseeing. Once, when Jack and I were returning after having both been away for two weeks together for scientific conferences, Miss Kit brought Alison, six months old at the time, to the airport to meet us. I put out my arms to hold her, but she didn't want me or Jack. She preferred Miss Kit. That was devastating, and I vowed to limit travel to as few days as possible. Another time, when the girls were older, they tallied up my travel time and claimed I was away for half the year. I protested and said that wasn't true. But they had marked calendars. It wasn't half the year, but it was pretty close; that was a second wake-up call to keep travel short and limited to only what was critical for my work.

The many conferences and workshops I attended were almost all-male affairs and, with few women in the field, most of my collaborators were also men. Years later, a (male) West Coast competitor still remembered me as "a practiced young lady" whose "angle was that she could convince many guys to collaborate." Not that I succeeded with all of them. When John Liston and I went to Chicago for a meeting of the American Society for Microbiology, we sat down for dinner with a leading microbiologist, Einar Leifson, who boomed across the table at me for all to hear, "Does your husband know where you are? Why aren't you home pregnant?" I had just published a paper on using the latest, most powerful electron microscope to study the structure of vibrios, but Leifson couldn't resist declaring, "Young lady, you cannot use electron microscopes to identify bacteria. You must use the human eye and brain to characterize them." He said the only way to look at bacteria was under an old-fashioned light microscope, using the staining technique that he'd developed.

Such carping wasn't as serious as losing a grant, though. I'd recently submitted a detailed grant proposal to the Food and Drug Administration (FDA) requesting $300,000 to $500,000—a lot of money at the time—to continue my work on microorganisms associated with seafood safety. The proposal received excellent reviews, and the FDA sent a four-man team for an on-site visit, common at the time. One of the four was John Liston, flown in from Seattle. Another came from the University of Massachusetts Amherst. I showed the men around my Georgetown laboratory, crowded with two graduate students, two undergraduates writing senior theses, and two technicians. I had not yet gotten permission to convert the ladies' room into more lab space. The team spent almost two full days with me. I made presentations, and they asked the usual questions about methods and work plans. Afterward, I thought it had gone well—until I got the news: the site team had turned down my grant request. Liston phoned to explain that I'd been blackballed by the professor from UMass Amherst. The rest of the team had argued with him, but couldn't get him to budge, and had no choice but to reject the proposal. The most crushing part was that Liston told me the man gave no reason for his veto; he just didn't like women scientists. In retrospect, I'm certain that blocking a young person's career advancement for personal reasons happened a lot. Geophysicist Marcia K. McNutt, the first female editor of the journal *Science* and the first female president of the National Academy of Sciences, remembers how, after a young woman who'd recently gotten a Harvard PhD gave a talk on the research she'd done for her thesis, the chair of the meeting session announced as soon as she'd finished her presentation, "We won't take any questions because everyone knows the results just presented are impossible, so we'll move to the next speaker."

There's no way I could have dealt with the injustice of my grant proposal's rejection by the FDA without Jack and our two wonderful children. We'd spend weekends as a family racing sailboats, hiking the Chesapeake and Ohio Canal towpath, and watching the eagles that nested across from our house along the Potomac River. Time with Jack and our daughters helped keep me calm—and happy.

This was the 1960s. I was the only woman in Georgetown's biology department, and the Washington, DC, area didn't offer many job alternatives for a professor of microbiology. Nearby Johns Hopkins University had only two women on its science faculty. The only way I could hold my own was to publish prolifically and demonstrate that my research was accurate and reproducible. Proving naysayers wrong is hard to do without funding.

By then, six graduate students and several undergraduate interns had joined my lab, so I hired a laboratory assistant: Janie Robinson, an African American woman ten years older and a foot taller than I was. Robinson had been our building's janitor. She'd become fascinated by the research we were doing and took to spending her off-hours with us. When I advertised for a lab assistant, she asked if she could apply. I told her of course.

Tutored by my laboratory manager, Betty Lovelace, Robinson proved to be a superb technician. She learned how to use staining dyes to reveal the shapes and identities of microorganisms; how to maintain cultures for years by growing them in test tubes and petri dishes of agar-agar or gelatin; and how to operate an autoclave, an instrument that combines high pressure and steam to sterilize instruments. Robinson became our gatekeeper and stayed with me for twenty years, until her retirement.

• •

There was one undercurrent at Georgetown that I did not like. In my very early years there, if I'd accepted a young woman as a PhD student, both my scientific reputation and that of the student would have been questioned. Ignoring the fact that I was also mentoring a group of young men, some of Georgetown's faculty would have assumed I was second-rate and could attract "only" women students. They would have similarly assumed the woman student wasn't good enough for a male faculty member to advise her research. The only woman graduate student in those early days whom I officially mentored was Minnie R. Sochard, who was already an accomplished—though overlooked—scientist. Sochard was an author of the *Atlas of Protein Sequence and Structure, 1954–65*, the first computer-based collection of protein sequences and a research tool used by many biologists. I also unofficially mentored Artrice Valentine Bader, a talented African American scientist who'd already done research at the NIH and was on leave to obtain her PhD. She was my age and a PhD student of George Chapman's, but I was a member of the committee that approved her thesis. I tried to convince her to accept a permanent post at Georgetown, but she preferred to stay at the NIH. To the university's credit, a large number of women were later accepted to the graduate program in our department and went on to distinguished careers at the NIH, other universities, and various companies.

In 1971, after seven years as a tenured faculty member at Georgetown, I was due for promotion to a full professorship, as was a fellow faculty member, who was also a Jesuit priest. I didn't expect trouble. Chapman had stated publicly at a faculty meeting that I was the most productive member of the department, and I'd become an associate professor with tenure early, three years after joining Georgetown. Grants are the universal currency of academic prestige, and I had more than $1 million in funding from the US Navy, the NSF, the NIH, and soon from the Environmental Protection Agency (EPA) for microbial ecology studies. I had data, research, and the requisite publications.

In short, I was utterly unprepared when Chapman called me to his office to say he had some unwelcome news. Promoting me was going to be impossible that year; the Jesuit priest would be promoted, and I would be recommended for promotion the next year. My hard work, talent, and new ideas backed by data—everything I thought would ensure my professional advancement—weren't enough.

In the meantime, more students were applying to work with me, and I had no space for them. A larger research department would offer more space and equipment than Georgetown. I decided it probably was best to move on.

Because of Jack's job, I had to stay in the Washington area. The region's most distinguished research university, Johns Hopkins University in Baltimore, was not yet coed and, like all the Ivy League schools in the 1960s, hired almost no women faculty. The University of Maryland, on the other hand, was nearby, coed, up-and-coming, had a historically strong microbiology department, and would be an ideal campus from which to conduct my research. I phoned a friendly colleague there and asked, "Do you have an opening at your main campus in College Park?"

"It's interesting you should call," he said. "We have a microbiology professor in your specialty, and he's retiring this year."

chapter three

It Takes a Sisterhood

Shortly before I moved, happily, to the University of Maryland, a highly qualified, part-time lecturer there was passed over for seven job openings. She asked a male colleague in clinical psychology why she hadn't been considered for even one of the positions.

"Let's face it," he told her, "you come on too strong for a woman."

As Bernice R. "Bunny" Sandler cried at home that evening, her husband wondered aloud, "Are there any strong *men* in the department?"

"They're all strong," she replied.

Then the problem wasn't Bunny, her lawyer husband said, "It's sex discrimination." If she were joining the department as the only strong person among a group of weak men, they'd be rejecting her strength. But if everyone is strong, they couldn't be objecting to that. They were objecting because she was a woman.

That simple statement—"It's sex discrimination"—began women in science's long journey toward banding together, arming ourselves with data, and tearing down the barriers we faced so stealthily that no one noticed what we were doing until it was too late to stop us. We'd be given a hand in this mission by Bunny Sandler and a woman in Congress.

• •

Congress had excluded white-collar professionals from the equal pay for equal work provisions of the Equal Pay Act of 1963 and the Civil

Rights Act of 1964. Three years later, however, President Lyndon B. Johnson issued Executive Order 11375 prohibiting federal contractors from discriminating against women when hiring. Researching the subject, Sandler spotted a footnote. "Being an academic," she said, "I read footnotes, so I turned to the back of the book to read it." And there, Sandler found a way forward for women like herself: by taking federal funds, the University of Maryland (and all research universities receiving federal funding) had become a federal contractor. "It was a genuine Eureka moment," Sandler later wrote. "I actually shrieked aloud."

Working with the Women's Equity Action League, a small Ohio group focusing on economic and legal issues with educational institutions, Sandler filed a class action suit against the University of Maryland and 250 other American colleges and universities. She based many of the cases on evidence of bias in hiring, tenure decisions, promotions, and salaries sent to her by women from all over the country. Although Sandler would win some of these lawsuits, the lack of enforcement powers in Johnson's executive order muted her victories. There were no penalties for the offending universities to pay or procedures put in place to reverse the inequities. Thankfully, the story didn't end there.

Instead, the issue of employment bias moved quietly to Congress, where Representative Edith Green, an Oregon Democrat and former schoolteacher, was passionately interested in women's equality in education. Having accumulated enough seniority to chair a House subcommittee on education, she held hearings on sex discrimination in academia. Then Green made a savvy move: she moved the legislation that would come to be known as Title IX through Congress so stealthily that almost no one—not the men who voted for it nor the educational establishment that would be tasked with carrying it out—knew what was happening.

Quietly organizing her forces, Green hired Sandler to coach her staff about gender discrimination in academia, while Vincent Macaluso, a sympathetic federal expert in the Department of Labor, provided secret advice on strategy. Morag Simchak, also from the Department of Labor, drafted an amendment so forbiddingly technical that she knew no one would even attempt to read it. Sandler reported tongue in cheek that Simchak, "having informed her superiors once about her work on drafting the amendment, did not feel any great need to inform them again." Several months later, Green, without informing her colleagues in the House of Representatives about the implications of what she was doing, slipped Simchak's short passage banning sex discrimination in any educational program receiving federal aid into what would become the Educational Amendments Act of 1972. Once the passage was in the bill, Green ordered the leaders of women's groups *not* to lobby for Title IX of the act. She didn't want to alert the men in Congress to its importance. And so, still happily oblivious to what Title IX meant, Congress made it illegal to discriminate against women in education and gave individual women the power to sue universities in order to hold on to their jobs, protest tenure or employment decisions, and demand equal pay.

The university athletic establishment considered the Educational Amendments Act of 1972 "a very minor bill," Sandler would later explain. The word "sports" was not even in it. A month later, Jonathan Spivak of the *Wall Street Journal* broke the real story under a masterfully understated headline: "Sex-Bias Section Could Spark Controversy."

Worried about losing federal funds, university lawyers promptly ordered some quick fixes. Old-style anti-nepotism bans against hiring faculty wives gave way to modern prohibitions against relatives supervising relatives. Almost overnight, a few leading scientists were leapfrogged three steps up the academic ladder, from research associate to professor, including two women mentioned in the previous chapter—Violet B. Haas, the brilliant mathematician in Purdue's engineering department, and Dora Priaulx Henry, the barnacle expert at the University of Washington—and several accomplished research associates at Harvard. Thousands of women suddenly received big pay raises, often without explanation and generally without back pay. For the first time, educational institutions had to publicly advertise job searches, end gender quotas for admission to graduate and professional schools, and pay equal stipends for male and female graduate students—although white men continued to get prestigious research assistantships while women and other groups underrepresented in science continued to receive less-prestigious teaching assistantships.

Above all, Title IX allowed us to believe change was possible. But those of us who also hoped Title IX would take care of sexism in science were soon dismayed.

••

In 1972, we had no idea how difficult it would be to change old attitudes. But over the coming decades, one scientist would be in a unique position to illustrate the depth and longevity of scientists' bias. That's because Ben A. Barres, the late chair of Stanford University School of Medicine's neurobiology department, was known as Barbara A. Barres for the first forty-three years of his life.

Raised as a girl, Barres felt from childhood that she was meant to be a boy. (Years later, Barres learned that her mother had been treated with a masculinizing drug to prevent miscarriage while pregnant with Barres, and that she had been born without a uterus or vagina, a condition known as Müllerian agenesis.) Too ashamed to talk to anyone about her confused feelings, she found solace in science. Her family was not wealthy and her parents hadn't attended college, but they lived in the New York area, so Barres could attend advanced science programs for young people at Rutgers, Columbia, Phillips Academy Andover, and Bell Laboratories. From these programs, Barres acquired an "intense and uncontrollable passion . . . for doing research."

As an undergraduate at MIT during the 1970s, when the university was still almost exclusively male, Barres was a science nerd like everyone else. And when a Nobel Prize–winning physicist made sexist remarks and showed nude pinups in class, she didn't complain; she just changed courses. Then she took a large artificial intelligence class composed almost exclusively of young men. One day, the professor announced in class that no one had solved a difficult mathematics problem he'd included on a take-home exam. Barres had solved it, though, and showed him her answer after class. Sneering, he accused Barres of cheating, insisting that a boyfriend must have solved it for her. "He just couldn't believe a woman had solved the problem when so many men had been unable to," Barres later said. (Forty years later, in 2017, a female freshman at a California state college had a similar experience when her math professor accused her of getting the correct answer to an exam question from the male student sitting next to her—despite the fact that the young man had himself gotten the problem wrong. She reported the incident to her teaching assistant, and the professor quietly retired at the end of the semester.)

Despite Barres's outstanding grades, no one on the almost exclusively male faculty at MIT offered her the university's customary research opportunities in their labs. As a graduate student at Harvard, Barres published six high-impact scientific papers but lost a prestigious fellowship competition, which would have helped her career, to a young man who'd published only one such paper. Then, one September day in 1997, Barres read an article in the *San Francisco Chronicle* called "A Self-Made Man," about an individual who'd transitioned from female to male. Barres had known it was possible to transition from male to female, but this was the first time she realized a person could do the opposite. Soon after, she began testosterone treatments, and was "flooded with feelings of relief." For the first time, she felt comfortable with herself.

After Barres transitioned, changing his name to Ben, he was uniquely able to gauge the true extent of the bias against women in science. He once heard a male scientist, unaware that Barbara Barres and Ben Barres were one and the same, say in passing, "Ben Barres gave a great seminar today . . . His work is much better than his sister's." Barres knew the scientific work he did as Ben was as good as the work he'd done as Barbara—but some people thought research performed by a Barbara couldn't possibly be as important. Barres also found that people treated him with more respect after his transition. "I can even complete a whole sentence without being interrupted by a man," he said. And in a remark that I still find particularly disturbing, Barres once said that "besides changing sex, the only time in my life I've taken an action that I thought might harm my career was when I decided to start fighting for the welfare of women in academia."

Sadly, Barres died of pancreatic cancer in 2017, at which point his life in science had spanned forty-five years of discrimination—all of them *after* Title IX, the law that was supposed to have solved our problems.

• •

The feminist movement was late in coming to science, but by the 1970s, women scientists were meeting privately in one another's homes to talk over their problems, far from their bosses' disapproving glances. Some were even writing memoirs about the injustices they had faced, although critics often called them aggressive and abrasive troublemakers. But we needed one another's stories to explain what we ourselves were experiencing—and to learn that we were not alone.

Despite Title IX, few men in power in the 1970s and '80s thought science, as an institution, needed wholesale reform. To meet the letter of the law just enough to keep their grants, men inaugurated what historian Margaret Rossiter called the Age of Revolving Doors, admitting exceptional women but then kicking them out as soon as the law and institutional rules allowed.

Biologist Sally Frost Mason entered graduate school at Purdue the first year Title IX was in effect. The biology department had admitted an extra-big class that year: half men, half women. All the women were assigned the same faculty advisor, a man who met with them individually and—the women discovered later—told each one the same thing: "You were admitted only because we were worried about losing our federal money." They didn't care whether the women succeeded or failed. Ignoring the insult, Mason earned a master's degree from Purdue in 1974 and began searching for a PhD thesis advisor there. She found one—and only one. And after he died unexpectedly, no one else in the department would take her on, and Mason had to leave Purdue to get her PhD elsewhere. But in a delightful story of sweet retribution, Mason returned to Purdue in 2001 as provost, the university's de facto COO. Later, as president of the University of Iowa, she raised more than $1 billion in donations, the gauge of success for American university presidents.

Another "revolving door" trick to avoid truly carrying out Title IX was hiring a token young woman as a relatively low-paid faculty member and making it impossible for her to advance. A talented woman named Lynn Caporale was offered an assistant professorship in biochemistry at Georgetown around the time I was told I'd have to wait one more year for a full professorship. Few men in her department had ever had a female colleague before, and as Caporale headed out to give her first lecture to three hundred medical students, a male professor offered her some "friendly" advice. "If you're nervous," he said, "why don't you wear a see-through blouse?" Caporale wasn't nervous. She'd lectured for large audiences before, and she certainly had no interest in doing a striptease at the lectern. So she asked her colleague sweetly, "Do *you* lecture in a see-through blouse?" Any woman would have understood the put-down, but the professor seemed genuinely mystified by her response.

Caporale won more grants than almost anyone else in her department, and her students awarded her Georgetown's Golden Apple for excellent teaching. Yet when she invited a colleague in her field to give a seminar, the university subsequently hired him at a salary $10,000 a year higher than hers. When time came to consider Caporale for a tenure-track position, she was rejected. When she asked why, a colleague told her privately, "There's just something that's not 'professor of biochemistry' about you." Caporale appealed the decision to her dean, who asked, "Well, if they don't want you, why would you want to join them?"—as if a university were a social club. Caporale got the message and accepted a job with far better pay at the pharmaceutical company Merck. But much the same thing happened there: like so many women, she was promoted into a dead-end office job rather than being given a spot in a research laboratory where she could have discovered new drugs and earned income from her patents.

University of Alaska oceanographer Rita Horner was targeted by, of all people, equal opportunity officers. When Title IX was two years old, the equal employment office in Fairbanks, Alaska, called her in and tried to cut her salary. Why, they asked, was she paid more than two male postdoctoral fellows in her department? "Because as an assistant professor, I outrank them," she replied. But the postdocs were married and supporting wives, her department chair noted. "You don't know who I'm supporting," Horner retorted. She won that battle, but when some of her laboratory space was given to a man, she gave up and left for the University of Washington, where she spent decades studying marine algae as a researcher in a nonfaculty position.

Bias could take many forms. Two years after Title IX came into effect, zoologist Sue V. Rosser was pregnant with her second child and working as a postdoctoral fellow at the University of Wisconsin. One day, the professor supervising her fellowship told her the birth of a second child would interfere with the lab's grant writing schedule. He told her to get an abortion. Rosser had her child, quit science, and became provost of San Francisco State University, making her one of the first women provosts at a research university. With experiences like these, is it any surprise that the number of women on prestigious faculties actually dropped during the 1970s?

Many women who did hold on were stuck in working environments rife with sexual harassment, due to the enormous power imbalance between established professors and the students they could flunk, fire, recommend, or hire—all practically on a whim. Physical closeness, late hours, and field trips added opportunities for sexual predation. I learned the term "sexual harassment" in 1976, a year after it was coined, when a female graduate student at the University of Maryland confided to me that female students warned each other to be wary of Professor X: "He gives you the goose. He pinches bottoms." Nothing ever happened to the man as a result of his behavior.

I would soon learn that universities, like other powerful institutions, know how to close ranks. In the 1980s, I was vice president of academic affairs at the University System of Maryland, the highest-ranking woman there. A group of female administrators, including two deans, asked for a meeting and told me that a department chair in the humanities was a serial predator who coerced students into sleeping with him by threatening to flunk them. As an official of a statewide system, I had no legal authority to investigate a case on one particular campus. And while there was eventually an internal investigation, the professor was allowed to take early retirement and keep his pension.

• •

As my colleagues and I began thinking about how to fight these kinds of injustices effectively, an episode involving an accomplished marine biologist nicknamed the Shark Lady showed that even the most talented and appreciated women scientists could be victims of discrimination.

When I moved to the University of Maryland, Eugenie Clark was one of a handful of female associate professors in science and mathematics. Departments at the time were very much entities unto themselves, so our paths didn't often cross. Clark was a gifted and glamorous

Japanese American ichthyologist who'd written international bestsellers—*Lady with a Spear* and *The Lady and the Sharks*—about her US Navy–funded research diving for poisonous fish in Micronesia. She pioneered the use of scuba gear for undersea research and discovered, by swimming among sharks, how they reproduce, sleep, breathe, and learn. A skilled fund-raiser, she raised money from a member of the Vanderbilt family to start the Mote Marine Laboratory & Aquarium in southwest Florida and helped found Egypt's first national park. One of the University of Maryland's presidents said Clark brought the campus more good publicity than the football team.

Yet as remarkable as Eugenie Clark was, she was forced to spend ten years as an associate professor, paid less than most male faculty members were paid at the start of their careers. When a group of women professors at Maryland banded together to agitate for better salaries, Clark's legal advisor warned her not to join them because the press would target her. "I can just see the headlines," he said, "'Shark Lady bites the hand that feeds her.'" Continuing without her, the women scored a major victory by getting themselves—and Clark—raises that brought their salaries almost level with those of their male counterparts. (Unfortunately, I was unaware of these women's efforts, as the microbiology department was located in a different part of the University of Maryland's campus.) Clark was later promoted to full professor.

There was also the problem that some of the most successful women in science were as biased as some of the men. As the only female scientists in their department, or sometimes even the entire university, they realistically viewed other women not as collaborators but as competitors for their token spot. For years, Helen Whiteley, whom you'll remember from the previous chapter, did not think other women scientists at the University of Washington needed support. Another woman, so powerful that her identity has never been revealed, expressed her disinterest in advocating for other women in science by saying, "It's not worth spending your time on losers. If they're any good, they'll make it." Women in the Endocrine Society worked hard to get Nobel Prize winner Rosalyn Yalow elected society president—but in her presidential address, Yalow made it clear that she wasn't interested in helping other women succeed, or even in thanking the women who'd helped her get elected. Instead, she announced that she found it "unfortunate that the women's caucus of the society has chosen to remain a special-interest group."

• •

Although we all felt less alone after telling one another our personal stories, I was eager to do more than just share sad and futile anecdotes. I wanted action—and other women felt the same way. But how could we fight to change a process we didn't understand? We realized we needed rigorous scientific data showing how sexism worked and how destructive it was to women's psyches and their bank accounts. Our strength would be achieving command of the facts.

Some activists had started meeting in the Washington, DC, area, and I joined them as often as I could. It was satisfying and a relief, really, to know that other women scientists felt as I did: the system had to be changed. At one of the meetings, I learned about a small survey that a young assistant professor, Alice S. Huang, was conducting at Harvard Medical School. Huang was born in China, but when she was ten years old, her parents had sent her to the United States, alone, to be educated at an Episcopalian boarding school for girls in Burlington, New Jersey. Her father was an Episcopal bishop, and St. Mary's Hall (now Doane Academy) prided itself on being the first school in the United States to offer girls an academic education equal to that given to boys. For three years, while Huang's parents remained in China, the school's principal was her legal guardian; Huang even lived with the principal and the principal's two sisters.

Alice Huang is a beautiful woman, and after graduating from Wellesley College, she helped pay her way through a PhD in microbiology at the Johns Hopkins University School of Medicine by modeling for a Baltimore department store. Huang soon became a rising star in animal virology (and along the way, she picked up an airplane pilot's license). She was eventually recruited to an assistant professorship at Harvard Medical School; soon after, fourteen research assistants and associates there began grumbling about their positions. Male professors asked Huang to investigate. It was a volunteer assignment on top of her regular duties, but she agreed.

Huang's 1972–1973 survey—while informal and unpublished—was

deeply unsettling. Almost all of Harvard's research associates and lecturers in biology and medicine were women, some of them more accomplished than Harvard's men. When Huang met the women one on one for lunch, some broke down and cried. No one had ever asked them about their experiences, and they'd never compared notes. "Women were afraid that, if we helped other women too obviously, we'd become tainted and suffer more discrimination," Huang later said. Her own female boss had ordered her not to hire any women.

Many of Harvard's research associates were victims of Harvard's anti-nepotism rules, which, until Title IX, banned hiring faculty wives. Four worked in their husband's laboratories, often multitasking as secretaries, technicians, dishwashers, assistants, and even laboratory managers. They worked without pay (or without decent pay), job security, pensions, sabbaticals, equal lab space, graduate students, or prestige. Divorce or widowhood could end a woman's career because when a professor left his laboratory, neither his research associates nor his wife inherited it. Other women Huang spoke to had remained single, although this did little to stop some men from openly looking down on them. In his book *From Dream to Discovery: On Being a Scientist*, Hans Selye, an eminent Canadian professor of psychology, described women scientists as "desiccated . . . bitter, hostile, bossy and unimaginative" and almost invariably in love with their boss.

Huang decided that women in biology needed to conduct a rigorously scientific survey—a census, really—showing who we were and what was happening to us. We knew we were far from alone in our troubles; many women in business, the arts, and government faced problems like ours. But we scientists had one advantage: we knew how to measure and document problems, and those of us with tenure could publish with relative impunity. And with a significant number of women scientists working in biology, it was both obvious and appropriate that useful information should be gleaned from this impressive cohort of women.

So Huang and three other female microbiologists—Eva Ruth Kashket at Harvard Medical School, Mary Louise Robbins at George Washington University's medical school, and Loretta Leive at the NIH—undertook the first statistically sophisticated, computerized study of the problems facing female PhD biologists over the course of their careers.

To its credit, the American Society for Microbiology agreed to finance the study, and a growing number of women met after work in one another's homes to help punch data into computer cards. In 1974, the research was published in the leading American scientific journal, *Science*.

The study showed that women advanced in their careers more slowly than men, that they were paid less at every stage, and that this salary gap expanded as men and women gained professional stature. For every dollar a man earned, a woman with the same degrees made do with, on average, 68 cents. While the vast majority of male scientists were married with kids, almost all faculty women were unmarried and had no children. Unexpectedly, the data showed that the male faculty treated single men with PhDs as deviants, keeping them in positions usually filled by women. "I wish I could tell you that study is out of date," Huang said in 2013, "but, unfortunately, its broad conclusions have stood the test of time." More recent studies have looked into the different and additional obstacles faced by women of color who pursue careers in science, technology, engineering, mathematics, and medicine (STEMM).

I did not expect that, by itself, one pioneering survey would convince many men. But to keep their data current and in the public eye, Huang and her allies developed a splendid idea. Everyone attending subsequent annual meetings of the American Society for Microbiology, from society members who'd just obtained their bachelor's degrees to full professors, was invited to pick up a colored thumbtack—blue for men, pink for women—and post their rank and salary anonymously on a giant graph posted in a hallway. Many men participated. And soon, passersby could see the pink and blue lines growing wider and wider apart. "That graph probably made people more aware than anything else," Huang later said. "No one had to give speeches, protest publicly, or say much of anything. People could just gawk."

Eager for advice on what else she could do to help women, Huang looked around for a female role model—only to realize she couldn't find many. "Those few I'd personally known never really got promoted to full professorship, and left the field after a few years," she realized. Still, she thought, perhaps Mary Bunting, the country's most powerful woman scientist, might have some advice.

Mary "Polly" Bunting was president of Radcliffe College and had founded the Radcliffe Institute for Independent Study to help women whose careers had been interrupted by family obligations return to work. Bunting understood firsthand the discrimination faced by married women trying to work in science. She earned a PhD in microbiology from the University of Wisconsin, but when she and her husband moved to Connecticut so he could take a position at Yale, she could only do her research as an assistant and lecturer.

Bunting was frank. "Alice, do not spend your time trying to help women now," she told Huang. "You have to focus on your own career. Only when you get into a position of power can you help women." In other words, we needed to be able to speak with authority before others would listen to us.

So Huang followed Bunting's advice. She concentrated on her career, including a stint as president of the American Society for Microbiology, and resumed working openly for women in science only after she became dean of science at New York University in 1991. When her husband, David Baltimore, became president of Caltech in 1997, Huang gave up her laboratory. But she went on to serve as president of the American Association for the Advancement of Science and became an active advocate for men and women sharing household and childcare responsibilities equally.

Huang soon recognized that Bunting had been right. She had thought that "screaming and showing people the data about our plight would make everyone change immediately." What she and other women activists came to realize was that "change would be gradual." By themselves, neither Title IX nor reams of data could change the way science was managed, but we continued to believe that if we could band together, we might get somewhere. And, in Huang's words, "if we could gain a little bit somewhere, we'd gain a little bit more the next time."

• •

If the only way to be taken seriously was to speak from a position of power, we knew we'd need a group of women scientists that could speak publicly in a single voice. And so, late one evening at the 1971 annual meeting of the Federation of American Societies for Experimental Biology, twenty-seven women stayed behind after the hotel wine bar closed to start the Association for Women in Science (AWIS). The organization would be open to women and their supporters across every field, although women in the medical sciences formed the group's core. Hematologist Judith Graham Pool and endocrinologist Neena B. Schwartz had tenure and felt secure enough in their careers to serve as co-presidents. Pool's discoveries about blood coagulation had already revolutionized the treatment of hemophilia and saved many lives. Schwartz had been hired at the University of Illinois College of Medicine in Chicago in 1953 because, as she told the story, "the only woman in the [physiology] department had become pregnant and the department Chair, George Wakerlin, felt it 'inappropriate for a woman in an advanced stage of pregnancy to lecture to medical students.'" One of Schwartz's most notable discoveries was the hormone inhibin, which is produced in the ovaries and inhibits the release of eggs. Men had spent years searching male laboratory animals for a hormone regulating the human menstrual cycle, but it was Schwartz who thought it was worth taking a look at female animals, too. Sure enough, she found the inhibin hormone. (Men's bodies were later found to produce small amounts of inhibin, too.) Despite the importance of her discovery, Schwartz dealt with discrimination for being a woman and for being Jewish, as well as battling her own fears about coming out as a lesbian.

One of the first questions AWIS asked was why women in the medical sciences weren't getting the grants they believed they deserved. AWIS soon discovered one big reason: women accounted for only 2 percent of the members of the advisory committees that recommended NIH grant requests for funding. A committee on breast cancer research had just two female members, and most committees had none at all. So AWIS threatened to sue the federal Department of Health, Education, and Welfare (which oversaw the NIH) for discrimination—and within months, the proportion of women on these important committees jumped from 2 percent to 20 percent. (The latest data also show that women do not apply for research grants as frequently as men; more analysis is needed to determine what other causes may be preventing women from receiving grants at the rate they merit.)

The successes Bunny Sandler and AWIS had in court made lawsuits look attractive to individual women scientists seeking recourse. But we soon came to realize that many of these suits dragged on for years, were decided in favor of the woman's employer, or were settled for pittances. Undiscouraged, AWIS has continued its advocacy for women.

• •

It was becoming increasingly clear that some of the biggest obstacles we faced originated in the professional societies to which we paid dues and for which we volunteered. In the late 1970s, for example, I was in the audience at a meeting of the Ecological Society of America when a colleague showed some inane slides of nearly naked women in provocative poses. During World War II, the military had used pinups to keep soldiers awake during training sessions—but among adult scientists? As an invited speaker, I'd been seated conspicuously front and center, and so I stood up and quietly left the auditorium in silent protest. But not one person followed me out or even said anything afterward—not that I'd expected anyone to. At the time, if a woman mentioned offensive behavior, the standard retort was, "Don't you have a sense of humor?"

Five years later, something similar happened at the annual meeting of the American Society for Microbiology (ASM). The ASM was (and still is) the world's largest organization of life scientists; at the time, it had 35,000 members, a third of them women. (In 2019, more than half of its members were women.) Robert P. Williams of Baylor College of Medicine, the society's president at the time, had a good reputation when it came to women's issues, and had convinced recalcitrant journal editors to publish the chart compiled by Huang and her collaborators to show the chasm between male and female salaries and ranks. Nevertheless, in his presidential address, he included a slide showing a cartoon drawing of a barely dressed young woman with two ice cream cones covering her breasts. "Things are not always as they appear," he quipped. Men in the audience snickered. Women sat in silence. But afterward, perhaps emboldened by the feminist movement, two women complained in a way that would have been almost unthinkable just five years earlier. Marjorie Crandall, a PhD mycologist at Harbor-UCLA Medical Center, and Louise Louden, a microbiologist in St. Louis, Missouri, wrote Williams an angry letter, asking him what his goal had been in showing the slide. Had it been "to make women microbiologists in the audience uncomfortable? Or to reinforce the fact that men are still in power and women are still sex objects? How," Crandall and Louden asked, "would you feel if a woman of distinction . . . show[ed] a slide of a scantily clad, attractive, young Caucasian man with two balls of ice cream covering his testicles [and] the ice cream cone covering his penis . . . ?" Unfortunately, Williams was more distressed about being called out for the slide than remorseful about including it in the first place.

The problem wasn't just seeing our bosses laugh at lewd images. As a volunteer on a score of different ASM committees, I could see men vigorously—even militantly—shoring up some of the biggest barriers to women's success. The ASM may have been just one roadblock keeping women out of biology, but it was a big and powerful one.

If everyone is to understand the events that subsequently unfolded, I'll need to explain briefly how the scientific system works. In the early 1980s, women earned 40 percent of the new PhDs in biology. After young people finished their PhDs, they spent two or more years as postdoctoral fellows, and those who wanted a career in academia would then seek out a post as an assistant professor as the first step in an academic career. They had six or seven pressure-packed years to prove themselves in this position in order to be promoted to associate professor, a position with tenure. Assistant professors who failed to measure up generally left their universities and found work elsewhere.

Success in science is measured by the talks you're invited to deliver, the grants you receive, the quality of your teaching, and especially by the number of articles you publish based on your own research. If you can't publish research findings in peer-reviewed journals, you essentially can't prove your worth. You won't be invited to talk about your findings, you won't be promoted, you won't be awarded grant money to do new research, and your graduate students won't get a shot at the best positions when it comes time for them to find jobs. If you can't publish your research to help advance the world's body of scientific knowledge, your research basically does not exist.

We knew men controlled access to the many important journals published by the ASM; journals were a revenue-generating enterprise for the organization, which had something of a monopoly on them. But we didn't realize the extent of the problem until, paging through the ASM's journals, we discovered that every single editor in chief was male. Below the editors in chief, 750 "expert" volunteers chose which articles to publish—more than 90 percent of those "experts" were men, and their decisions were subject to neither challenge nor review. (Almost twenty years later, women in the movie industry would publish similar data about their experiences. On films directed by men, women made up 11 percent of writers and 21 percent of editors. But on films with at least one female director, women accounted for 72 percent of writers and 45 percent of editors.)

In short, women scientists were paying membership fees to the ASM, and the organization was failing to meet its responsibilities to us in return. We needed to understand the ASM's power structure to figure out how policies permitting such gender imbalances were made.

In the 1980s, the ASM was run by volunteers. Meeting in the home of Sara W. Rothman, a microbiologist at the Walter Reed Army Institute of Research and a leader of our coup, we spread big sheets of paper on the dining room floor and mapped the labyrinthine ways those volunteers made decisions. No matter the issue, the routes kept leading back to the president of the ASM. Each president served only one year, but he (invariably "he") could sit on powerful committees for three years: one year as president-elect, one year as president, and the following year as immediate past president. To change the ASM, we had to gain access to the presidency.

Since its founding in 1899, the ASM had had only three women presidents, roughly one per generation. The first was Alice C. Evans in 1928; Evans, as I'd learned during college, had discovered that unpasteurized milk could transmit a debilitating and even fatal disease. The second, Rebecca Craighill Lancefield, was the leading authority on another deadly bacterium, *Streptococcus*. Her 1943 presidency was widely and insultingly attributed to "all the men being away" during World War II. The third, in 1975, was Helen Whiteley from the University of Washington, who'd discovered a bacterium that could be used to make cotton and tobacco crops insect-resistant. The ASM was filled with talented women who would have made great candidates to succeed Whiteley, and the playing field had to be leveled so they could have the opportunity to do so.

Fortunately, we had two prominent male microbiologists on our side.

When Albert Balows became president in 1981, he appointed women to half the slots on the ASM's presidential nominating committee. A year later, Frederick C. Neidhardt of the University of Michigan announced that a key goal of his presidency would be expanding the ASM's top ranks to include more women. Without telling me, Neidhardt decided I should run in the 1983 election. We didn't know each other at the time, but Neidhardt later said he somehow knew I felt "very, very strongly" about increasing the role of women in the ASM. He was right—I did. Still, it was a surprise to learn that I had been nominated. I was delighted, and while I felt some trepidation, I was honored to take on the challenge. (When the ASM's Committee on the Status of Women in Microbiology created the Alice C. Evans Award to honor members who helped women in the profession, Fred Neidhardt was the first recipient.)

Among the biggest obstacles to women holding office in the early 1980s was the pervasive doubt that we could be effective leaders. I ran against two male candidates on a businesslike, gender-neutral platform that I believed would appeal to the (male) majority of ASM members. Carefully avoiding inflammatory words like "women" and "female," I campaigned for "electronic mail" in ASM headquarters, better financial planning, committee appointments for "younger members" (code for women and minorities), and greater recognition for "clinical microbiologists" (code for female hospital laboratory technicians)—and won the election.

As president, I could launch a few small but long-overdue changes: creating travel fellowships to help young scientists and lab technicians attend ASM meetings, and locating inexpensive housing (generally in college dormitories) for those who could not pay hotel prices. I made funds available to women and underserved groups—including African American, Latinx, and other minority members—so they could host recruitment receptions. And we arranged for there to be childcare facilities at the society's annual meeting.

These changes were steps forward, but to enact long-term solutions, we needed women to hold the office more than a single year every few decades. But to my dismay, when I chaired the ASM's nominating committee the year following my presidency, the next official slate of nominees consisted of two male candidates whose priorities did not appear to include helping women. If we were going back to business as usual, another quarter century might pass before another woman president was elected.

No one had the power to overrule an announced slate of candidates. That would have required a change in the society's bylaws, and I'd already learned that one of the most important rules of effective leadership is "Don't start a battle you have no chance of winning." Women were still in the minority in the ASM, and if the male members united against us, we'd lose. So I took it upon myself to understand the arcane rules of the nominating process; if we knew the rules, perhaps we could figure out a quiet and nonconfrontational way around them.

I did know one way to skirt the official nomination process. The year I ran for president, the most outspoken critic of my scientific work, Richard Finkelstein, a medical microbiologist at the University of Missouri, ran against me as a write-in candidate. That got me thinking about how men had won the presidency for eighty-two of the ASM's eighty-six years in existence. Couldn't women have just written themselves in? It took a while, but I finally had one of those *aha* moments when all becomes clear. The reason women became presidents once a generation was so simple, so effective, and so obvious that I'd never noticed it before: the ASM mailed its newsletter telling members how to launch a write-in campaign well *after* the deadline for write-ins had passed. By the time the newsletters arrived, the official slate was set in stone. Only insiders like Finkelstein, who'd served in several official capacities within the ASM, would know the names of candidates in time to propose alternates. The system had been foolproof for decades.

The deadline for write-in candidates was only a week or two away, so time was short. Modifying an organization so set in its ways was going to take creative thinking, diplomacy, and help. Fortunately, we were ready.

I phoned Anne Morris-Hooke, who was then at the Georgetown University School of Medicine. Born in Australia, Morris-Hooke was a passionate opera fan who tracked singers the way foodies track restaurant chefs; her favorite was the Australian soprano Joan Sutherland. Morris-Hooke had received her graduate degrees from Georgetown and was a spirited member of the ASM's Committee on the Status of Women in Microbiology.

After swearing Morris-Hooke to secrecy, I explained that if another woman was to become president any time soon, we needed our own write-in campaign. But that wasn't all. To override the society's slate, the women of the ASM would have to put their support behind *one* write-in candidate. We'd never win if we split our votes. We would have to be highly organized, focused, and quiet—which is why this is the first time anyone involved in our insurrection has revealed the full story of what happened.

The need for secrecy must seem strange today. But until recently—especially before the #MeToo movement—women always had to work behind the scenes. If large numbers of men (and even certain women) had known what we were doing, I'd have made so many enemies that I would never have been able to accomplish anything in the future. "Just joining a feminist group was poison, the kiss of death," Morris-Hooke recalled. "You would be talked about if you attended feminist groups." Sara Rothman recalled that, "If two women stood at an elevator, a man would come by and say, 'What are you two ladies plotting?'" Her boss at the Walter Reed Army Institute of Research complained about her spending time with "*those women.*"

I could privately launch and advise the campaign, but Morris-Hooke would be its spunky operational director—and we needed a woman who was both a well-recognized scientist and an experienced manager to agree to run as our write-in. Jean E. Brenchley, a leader in bacterial metabolism and regulation who was moving to Penn State University at the time, courageously volunteered. We had only a week to amass at least fifty signatures—one hundred would be safer—to nominate her as a write-in candidate. That sounds easy today, but we barely had fax machines in 1985.

Over the phone, I recommended some formal language so Morris-Hooke could have petitions duplicated and handed around: "We the undersigned members of the ASM hereby nominate Jean E. Brenchley to be a presidential candidate." Nothing in the petition would mention anyone other than Brenchley by name. "We stayed quietly in the shadows," Rothman remembered, "totally energized and giddy."

Collecting signatures in Washington-area government laboratories turned out to be easier than we'd anticipated. Before the week was out, more than one hundred members had signed our petition, which we delivered to the ASM right on deadline. Fred Neidhardt, who'd suggested me as a candidate in 1983, published an impassioned plea in ASM's newsletter exhorting members of both sexes to vote for women.

Jean Brenchley became the first woman write-in candidate in ASM history to be elected president.

Her election enraged several men. Former ASM president John Sherris was miffed that Brenchley had run against his committee's official slate and wrote to the Committee on the Status of Women in Microbiology about his "concern" that the write-in process could be "used to further the interests of particular groups within the Society." (There's no record of Sherris complaining about Finkelstein's write-in candidacy two years before.) Replying tartly to Sherris's letter, Viola Mae Young-Horvath, the chair of the women's group, predicted that Brenchley would "make every effort to represent equally *all* members of the society . . . [That this] has not always been the case in the past is a well-documented fact."

Outgoing president Moselio Schaechter summoned the Committee on the Status of Women in Microbiology to his elegant suite at ASM's general meeting. (ASM presidents normally stayed in the convention hotel's large presidential suite, although I did not. When elected, I was told the rooms were needed by the outgoing president.) The committee had done "a very dangerous thing," Schaechter said, and it could cause "many problems for the Society."

Tempers cooled slowly. Barbara H. Iglewski, an eminent scientist and department chair at the University of Rochester Medical School, was nominated as a candidate to succeed Brenchley. Meeting Iglewski on a sidewalk outside an ASM meeting, a man who'd expected to be nominated instead of her grabbed her by the shoulder and asked angrily, "What are you women doing? How are you selecting these nominees at the expense of far more qualified men?" Iglewski won the election.

In the dozen years after my presidency, six women, including Brenchley and Iglewski, became presidents of ASM—more women presidents than the society had ever had before. The first three held the office one right after another, instead of twenty years apart. And except for Jean Brenchley, none ran as write-in candidates—they were all chosen by ASM's nominating committee. (Today, women are elected president roughly proportionally to ASM's membership. Not bad for an insurrection.)

But much remained to be done. We hadn't become ASM presidents "for the glory of it," the reason one man had given for running for office. We wanted to permanently democratize the association. To do that, we would have to apply sustained pressure for many years to come.

• •

Shortly before Barbara Iglewski became president in 1987, I urged her, "If you really want to do something for women after you're president, join one of the committees where the term is long enough for you to achieve something significant." Iglewski must have, because she spent nine years, from 1990 to 1999, as an extremely effective publications chair. Remember how the editor in chief of every single journal published by the society was male? And how 90 percent of the reviewers who screened articles for publication in those journals were men? Well, Iglewski lobbied for more women editors and editorial board reviewers. When Alison D. O'Brien became editor in chief of the journal *Infection and Immunity* in 1999, we all celebrated. Ten years later, O'Brien was elected president of the ASM.

As part of our long-term campaign, three of us worker bees also managed to be appointed to the ASM's powerful governing board, the Council Policy Committee. Iglewski's role as publications chair gave her a seat on the board, while Anne Morris-Hooke became secretary of the ASM (as a write-in candidate, incidentally), and I was elected chair of the American Academy of Microbiology. As chair, I reorganized the AAM into a prestigious honorary society for leading microbiologists and increased the number of women in its formerly male-dominated membership. In the end, all members of the ASM benefited from the inclusion of women in the society's leadership structure. But the larger battle continued.

• •

Consider the case of Samantha "Mandy" Joye, who came from a struggling farming family and was bullied throughout her childhood, but who doesn't take garbage from anyone. In the mid-1990s, she went to a very well-known male scientist's office to interview for a research position. "He had a *Playboy* bunny calendar on the wall and an African fertility god sculpture, a giant penis, on his desk," she recalled. "The penis was right in my line of vision, right in front of me. I was so exasperated I finally said, 'You have to move that thing.' "

He said, "What thing?"

"The sculpture," Joye said. "Get it out of my face."

Joye decided to go elsewhere for her postdoc and became an oceanographer at the University of Georgia. "I think it's pathetic that today, I still have to worry about how my daughters are going to get treated when they're in high school and college," she recently said. "You have to have a tough skin to be a scientist, but there's no reason for inappropriate, asinine behavior."

Consider this story, too: In 1999, several years after Joye applied for that research position, Marjorie M. "Kelly" Cowan, an assistant professor at Miami University in Ohio, attended a party at an annual meeting of the ASM in Chicago. When a prominent member of the organization offered her a ride back to the hotel in his limo, she thanked him and climbed in. He was married and in his seventies; she was in her thirties. Within thirty seconds, he'd leaned over, grabbed her, kissed her on the mouth, and felt her breasts. As she recalled years later, her mind froze for maybe half a minute before she recovered from the shock and pushed him away. Later, back in her hotel room, she tried to rationalize his behavior. "Maybe he'd had too much to drink, or maybe he was having a dementia episode," she told herself. When he tried again at another meeting, she asked him, "Aren't you married?"

"Yeah," he replied. "I love my wife."

Several women in the ASM complained to management about his behavior, but did he lose his university job? No.

So the battle continued.

Our next engagement would occur in a prestigious, predominantly male engineering school, where a fight would start over tropical fish tanks.

chapter four

The Power of Sunlight

A Radcliffe College undergraduate from the class of 1964 was doing advanced scientific work in a Harvard laboratory, analyzing data at her desk, when the door flew open.

"There stood a scientist I didn't know 'personally' but recognized instantly," the undergraduate, Nancy Hopkins, revealed publicly for the first time fifty years later. "Before I could rise and shake hands, he had zoomed across the room, stood behind me, put his hands on my breasts, and said, 'What are you working on?'" Hopkins didn't know which embarrassed her more: that the man was Francis Crick, the godlike co-discoverer of the structure of DNA, or that his behavior so demeaned him.

Even today, after a career as an MIT professor and a member of the National Academy of Sciences, Hopkins sometimes regrets having shared this story at all. Crick was visiting James Watson. The men were close; they'd shared the Nobel Prize after using Rosalind Franklin's X-ray photograph to elucidate the structure of DNA without her knowledge or consent. Watson was Hopkins's mentor and close friend. He was the man who'd told her she had to get a PhD, a buddy who'd never laid a hand on her or made a pass at her. If she complained about Crick's behavior, she'd embarrass Watson and spoil the party she'd been invited to that night. Besides, what could she do? Women knew that men looked at them as sex objects; that was just life.

"What I did not grasp till years later," Hopkins said, "was that a man who treats a student that way may not be genuinely interested in her lab notes."

It would take Hopkins three decades to face the fact that she worked inside a sexist system, and that her talents, hard work, and groundbreaking experiments would not be sufficient for her to win out in the end. Even after she'd gathered the data to prove the problem was gender discrimination, she still wasn't convinced.

As scientists, we always knew data were important. The question was what to do with it. Her colleagues helped Hopkins answer that question—but it was only when they orchestrated an emotional confrontation with the administration and spread the facts around the academic community that their power truly shone.

Though much later, Hopkins would pay a price—a heavy, personal price—for her courage.

• •

The story of this rebellion starts when MIT recruited Nancy Hopkins to join its faculty in 1973, a decade after her encounter with Francis Crick in that Harvard lab. Title IX had just banned discrimination against women in educational institutions receiving federal aid. MIT was especially vulnerable: women accounted for more than a third of its undergraduates, but only 8 percent of its faculty—and, with only a small endowment, MIT could not risk losing its federal grants. Hopkins had had superb training in molecular biology as a Harvard graduate student, and her research was much talked about. So Hopkins became a faculty member at MIT. She considered herself a beneficiary of affirmative action, "approximately number 10" of the university's female hires after the passage of Title IX.

Almost as soon as Hopkins began working at MIT, a woman administrator warned her that the university's female students were having problems because professors wanted to date them. "I didn't understand immediately the seriousness of this problem," Hopkins later admitted. "Put men and women together in the workplace, what did they expect?"

A few years later, I had my own opportunity to observe something worrisome at MIT when I served on a committee to review the university's accreditation. Part of our charge was to report on the relationship between the university, its students, and its faculty. At the request of the men on the committee, two female committee members and I privately interviewed female undergraduates, graduate students, postdoctoral fellows, and faculty members, and reported on what we termed the "special vulnerability—intellectually, emotionally, even sexually—of these women to men [who were] often . . . prejudiced, manipulative and unsympathetic to their requirements and concerns." Students confided to us that some male classmates and faculty patronized them. The longer women stayed at MIT, the lower their aspirations became. I remember one student in particular who, when asked what she hoped to do after graduating, "guessed she could get a job using her MIT PhD to design children's toys." Women faculty members talked about low salaries and insufficient laboratory space. When we told Provost Walter A. Rosenblith that MIT's women felt they weren't being treated very well, he replied that this disturbed him greatly. A man of sincere and deep compassion, he looked as if he might cry.

Hopkins was not yet a feminist in 1973. She thought the federal government and the women's movement had eliminated gender discrimination for her generation. The problem, as she saw it, was that "high-level science—the only type I was interested in—required that you work 70+ hours a week. How could you possibly be the kind of scientist I aspired to be and be a mother?" The belief that a mother couldn't possibly be a great scientist was so strong that more than half the tenured women at MIT had no children.

Hopkins was married and had been planning to have children before she turned thirty, as soon as she finished her postdoctoral fellowship; amniocentesis and in vitro fertilization were not yet available to help older women have children. But Hopkins and her husband divorced, and she subsequently decided not to remarry or have children. "Looking back," Hopkins reflected in 2015, "it's hard to understand how I could have been *quite* so slow to recognize that a profession in which half the population can't participate equally and also have children is *by definition* discriminatory." It would be years before she saw that "the way science careers and institutions are structured is an artificial and hence changeable system designed by men, for men, in an era when men had full-time wives to care for their families."

As the years passed, she attributed any lack of success in her career to her "own failings, particularly not being sufficiently aggressive or self-promoting in a highly competitive profession." "My response," she later said, "was always to work harder, to try to do a better experiment, on the theory that if you did a Nobel Prize–winning experiment, you wouldn't *have* to be self-promoting—everyone would *have* to acknowledge your discovery."

One day, Hopkins heard that Christiane Nüsslein-Volhard, the world's leading developmental biologist, was switching from studying the genetics of fruit flies to the genetics of a vertebrate. Hopkins was interested in the genetics of behavior, so she took a sabbatical and traveled to the University of Tübingen in Germany, where Nüsslein-Volhard was exploring the effects of genetic changes on offspring. Nüsslein-Volhard had just opened a 6,000-tank fish house for 100,000 *Danio rerio* (zebrafish), a tropical fish popular in home aquariums. Zebrafish are so transparent that light passes through them, so you can watch their organs form and grow within them.

Almost immediately, Hopkins fell in love with zebrafish and decided she would try to use them to study the genes responsible for the early development of vertebrates. It would be risky. The technology she'd need did not exist; if she succeeded, it would be a miracle. But she was determined to try—and to start, she'd need more lab space for fish tanks. Her MIT laboratory barely had enough room for a handful of graduate students.

So in 1993, Hopkins went to her department chair and asked for more space. For a tenured senior professor like Hopkins, this should have been a routine request; it wasn't difficult to assign a faculty member additional space. But the chair refused point-blank to give her any extra room. Instead, he told her that "only second-rate people go into zebrafish." First-rate geneticists stayed with fruit flies.

"But it's a new science," Hopkins protested. Nüsslein-Volhard was doing it.

"How do you spell that name?" the department chair asked—he didn't know who she was. (Just three years later, Nüsslein-Volhard would win a Nobel Prize; she used some of her prize money to help women scientists cover the costs of housekeeping, cooking, and childcare.)

Hopkins had already begun to wonder whether men took her work seriously. When a senior scientist who'd praised reviews she'd written asked her to sleep with him, what disturbed her most was the thought that her work might not be as good as he'd said it was. A new genetics course she'd developed was taken over by male professors, who planned to use it as the basis for their own book. According to an article in *Science*, Hopkins "stopped teaching entirely in protest." Another department chair she greatly respected said she couldn't teach a different genetics course because male undergraduates "would not believe scientific information spoken by a woman." The worst part was that she knew he was right.

But the final straw was not getting space for her fish tanks.

Hopkins found herself going to work each day accompanied by "tremendous bitterness, hopelessness, despair, sadness, and the belief no one would ever understand." And yet "there was no one to tell about it. No one would believe you. Women in my generation—we had to think we were crazy." Was she really being treated unfairly? she wondered. Were MIT men really getting more laboratory space than she was? To a scientist like Hopkins, the way to find out was obvious: measure the laboratories.

Hopkins didn't want to make enemies. "When I went to MIT department meetings," she recalled, "there'd be people who'd had a fight 30 years before and still hated each other." She was afraid of being seen as a complainer and a troublemaker—as a bad scientist. And yet she had to know. Fortunately, these were the years before there was such intense focus on turning scientific discoveries into profitable patents, and most laboratories were collaborative and welcoming to fellow researchers. So, armed with her tape measure, she went from lab to lab in MIT's School of Science, recording the size of each. It took her a year, but she got the facts she needed. She was right: the men had bigger labs—some four times larger than her own. The labs of senior male professors averaged 3,000 square feet, while senior female professors had, on average, 2,000 square feet in which to work—roughly as much as junior male faculty members.

Hopkins was furious. She gathered up the pile of memos, complaints, and letters about laboratory measurements on her windowsill and marched them to a lawyer, who identified the problem: discrimination. Hopkins started thinking about suing MIT. She decided that she "couldn't do science unless something changed."

During the summer of 1994, Hopkins drafted an angry letter to MIT president Charles M. Vest. At the last minute, she decided to get another woman's opinion before sending it. She invited Professor Mary-Lou Pardue to a local café for lunch and carefully spread the letter out on their table for Pardue to read. To Hopkins's surprise, Pardue, who'd been elected to the National Academy of Sciences eleven years earlier, immediately cosigned it. Neither had ever confided in each other or in their colleagues about their frustrations; they weren't even sure how many female colleagues they had. "Check the back of the [course] catalogue," Hopkins suggested. "Maybe they list the women separately?" (They didn't.) Pardue and Hopkins were startled to discover that MIT's six science departments (biology; mathematics; physics; chemistry; earth, atmospheric, and planetary sciences; and brain and cognitive sciences) employed 197 tenured men and only 15 tenured women (including themselves); two more women were assigned primarily to MIT's School of Engineering. When Hopkins and Pardue showed the letter to the fifteen women, all but one—who said she'd faced no discrimination—instantly signed. Within twenty-four hours, they had become a small but united group of women.

On average, these women were more distinguished than the men in the School of Science; 40 percent were members of the National Academy of Sciences and/or the American Academy of Arts and Sciences. Twenty years later, the breakdown was much the same: Of the 16 senior women in MIT's natural sciences and engineering departments who had signed Hopkins's letter, 4 had been awarded the National Medal of Science, compared to 7 of the 162 male full professors in those departments. And 11 of the women would be elected members of the National Academies of Sciences, Engineering, and Medicine, compared to 11 of the 162 male professors.

Still, Hopkins and her colleagues did not want to be viewed as troublemakers, so they decided to work in secret. In August, they made an appointment with Robert J. Birgeneau, MIT's dean of science, to hand-deliver Hopkins's letter to him. One by one, each woman who'd been able to join the meeting told her story. One of them described her career as "death by a thousand pinpricks."

For Birgeneau, a warmhearted man and extremely effective administrator, the meeting was "simply overwhelming . . . akin to a religious experience," he later told a reporter from *Science*. He knew these women were among the nation's leading scientists, but he hadn't realized how distressed they were. Birgeneau said that if just one woman had come to him with those complaints, he might have written them off as a personal problem with her boss. But hearing fifteen stories from the fifteen distinguished women present in the room, he decided Hopkins was right: this was discrimination. MIT had a legal problem with Title IX, but the bigger issue was human—and systemic. The higher these women rose in their profession, the more marginalized they felt. The university needed to address this problem for the sake of its professors and itself.

Leaving Birgeneau's office, the women felt as if they were floating and danced down the street.

• •

In an era when most universities were still ignoring complaints from their women faculty, Robert Birgeneau formed a confidential committee to gather more data—with MIT president Charles Vest's strong support. (Vest had grown up in West Virginia and was sensitive to injustice wherever it occurred.) At first, Birgeneau worried Hopkins was "too radical" to chair the committee, but none of the other women would take her place. His biggest challenge proved to be the university's male department chairs. Most of them opposed the committee altogether; Hopkins recalled a September 1994 meeting where the men "just sat there looking stony" until Birgeneau compromised by appointing three men to the hitherto all-female group. Two of the three men, including Nobel Prize–winning physicist and humanist Jerome Friedman, quickly allied with the women.

The committee's message was "data-driven," Birgeneau later said, which was "a very M.I.T. thing." And the data were stark. While more than half the undergraduates in three of the School of Science's six departments were women and the number of female PhD students was rising nationwide, MIT's percentage of female science faculty had been stuck at about 8 percent for twenty years. Women were getting lower salaries, pensions, and funds to start their labs; less equipment and heavier teaching loads; and fewer nominations from MIT for awards, department chairmanships, and influential committee seats. When MIT professors got outside job offers, MIT made counteroffers to the men, but not the women. There had never been a female department head in science or engineering. Senior female faculty felt marginalized and powerless within their departments. Junior female faculty were relatively content, but said their main difficulty was balancing career and family; more than half the tenured female faculty had no children.

Birgeneau began correcting problems even before the committee finished its report. Laboratory space and salaries were reasonably straightforward to address. Hopkins got 5,000 square feet for her fish tanks and was freed from teaching so she could devote thirty to forty hours a week to her research, as well as women's equity issues. Some women received raises of 10 percent in one year. To increase the number of women faculty, Birgeneau refused to approve male hires until all the women on the candidate list had also been interviewed. He recruited so many women that they show up on later graphs as "the Birgeneau bump." Thoughts of a lawsuit disappeared; Hopkins and the women on the committee wanted to do their science in peace.

But how had such inequities come to exist in the first place? The committee's report concluded that "most discrimination at MIT, whether practiced by men or women, is largely unconscious." This was news for most scientists, although psychologists had been documenting over and over again since the 1970s that both men and women unconsciously *over*value work they think was done by a man and *under*value work they think was done by a woman. Today, Hopkins says the discrimination happened because "the men were running the darned thing. The women were invisible, chugging along writing more grants and raising money. It just happened, and nobody was thinking to look for it."

The committee delivered its confidential 150-page report in 1996. Parts of the report were sent to relevant department heads, but the full document was read by only three people: President Vest, Provost Robert A. Brown, and Birgeneau.

For the next three years, the committee debated how much of the report should be made public. Finally, in 1999, five years after Nancy Hopkins's initial request for more lab space was rejected, the report was published in the MIT faculty newsletter, with President Vest's enthusiastic endorsement. "I have always believed that contemporary gender discrimination within universities is part reality and part perception," he wrote in an introduction, "but I now understand that reality is by far the greater part of the balance."

When women reporters heard that MIT, the nation's premier science and engineering university, had admitted to having systematically deprived distinguished women scientists of their fair share of resources, they leapt at the opportunity to cover the story. The *Boston Globe* ran reporter Kate Zernike's story in its Sunday edition under the headline "MIT Women Win a Fight Against Bias. In a Rare Move, School Admits Discrimination." Two days later, the *New York Times* published Carey Goldberg's "MIT Admits Discrimination Against Female Professors." Many readers had probably not given much thought to women scientists before, much less the discrimination they faced. Without these articles and others like them, Hopkins believes the report would have been quickly forgotten, as a 1983 report compiled by MIT graduate students in computer science had. This time, hundreds of female scientists who had read the report or the articles covering it emailed Vest to share their own experiences.

Reading Vest's statement, Hopkins felt her bitterness and pain begin to slip away. "The idea that powerful people in the university had heard us and said, 'Yes, you're right'—to me that was the high point of it all," she later said. When Hopkins arrived at work the following week, the hallway outside her office was clogged with TV camera crews, and inside, her phone was ringing. When she picked up, a voice said, "You're on the air. This is Radio Australia."

Hopkins, who'd entered the workforce at a time when women worried that speaking out would destroy their careers, was invited to lecture about the "MIT Miracle" at more than a hundred campuses across the country. In the White House, Bill and Hillary Clinton read the report and asked MIT to send Birgeneau and a faculty member to a national Equal Pay Day celebration on April 7. "It has to be you," Vest told Hopkins. At the event, she was seated next to Hillary Clinton, two seats away from President Clinton, facing an audience and a wall of cameras. She was so nervous, she couldn't remember what to call a president. "I think I called him 'Mr. Clinton,'" she says. Both Clintons gave short speeches praising MIT and extolling the importance of women to American science and the US economy. *Science* and the *Chronicle of Higher Education* wrote major stories about the event (although the *Chronicle of Higher Education* referred to Hopkins as "Mr. Hopkins" three times).

There was pushback, of course. A *Wall Street Journal* editorial denounced the MIT report as "politicized exercises in 'social science,'" arguing that women faculty lowered the university's high standards and that the committee (composed of some of the nation's finest scientists, including a Nobel Prize winner and several members of the National Academy of Sciences) had not used proper "scientific process" to evaluate the professors' complaints.

Vest and Birgeneau wrote back in protest: "First, we were told that college women just do not 'want' to do sports. Then because of Title IX, the most significant sports event of 1999 was the US women's soccer team winning the World Cup. Next, we were told that college age women just do not 'want' to study science. But now, at MIT, more than 50 percent of our undergraduate science majors are women and the only characteristic distinguishing them academically from their male fellow students is that they have higher graduation rates. Now, the *Wall Street Journal* tells us that women just do not 'want' to be college science professors." Their point was clear: women were not underrepresented in science by choice.

Over the next decade, the committee's report about MIT's School of Science became a model for reforming other parts of MIT and other universities and for examining bias against female, African American, and Latinx scientists. Vest invited the presidents of eight other universities to come to MIT and commit to doing similar studies to eliminate gender bias. MIT became one of the few universities in the US where men and women earned the same salaries and women became department chairs. In 2004, neuroscientist Susan Hockfield became its first woman president. Perhaps most important of all, MIT decided that "no faculty member—male or female—should be disadvantaged by family responsibilities." By 2014, thanks to new family leave policies, automatic extensions if a woman has a child before getting tenure, new on-campus daycare centers, and subsidies for childcare while traveling on business, Hopkins could report that the new norm was for junior women faculty to have children.

Far from being sidelined, the women who feared they'd be branded as troublemakers for speaking out had thrived. "My research blossomed," one woman commented, "and my funding tripled. Now I love every aspect of my job. It is hard to understand how I survived those years—or why." Hopkins's career took off, too. Her laboratory was expanded to make room for two dozen graduate students and 150,000 zebrafish, and her team identified at least 25 percent of the genes a zebrafish needs for early development. Thanks to her innovative research, Hopkins was awarded an endowed chair at MIT and elected as a member of the National Academy of Sciences.

• •

I believe the MIT report was a seminal moment. The fact that its findings were carefully documented—and accepted by the president of the leading scientific university in the US—affirmed what women had been saying all along. The report did for gender discrimination in universities what the Anita Hill–Clarence Thomas hearings had done for sexual harassment in 1991.

Marcia K. McNutt, president of the National Academy of Sciences, believes the report was the beginning of the end of most overt discrimination toward women in science. Since then, she says, girls are no longer told, "No, no, no. You shouldn't be here, you can't make it, you will harm the field."

How did the revolt begun by Nancy Hopkins start to succeed? Robert Birgeneau believed the committee's study evaded the dustbin not because of its legal ramifications, but because of the power of the group behind it. Oceanographer Sallie W. "Penny" Chisholm, who served on the committee, says it succeeded because the women suspended personal suspicions about one another, overlooked differences among their disciplines, and concentrated instead on their common experiences. I think the MIT revolution worked because, for the first time in academic science, women rose up together to press for change. The #MeToo movement has given a new generation the same lesson: when women work toward a shared goal, they are a formidable force.

"Teach girls to stand up for themselves," molecular biologist and former president of Princeton University Shirley Tilghman said. "If the world were completely fair, we wouldn't have to do that but, as of right now, it is not fair—so we do."

Nancy Hopkins agreed, but also emphasized the need for finding allies in the right places. "I think perhaps the most important thing we learned [from the report]—and have relearned many times since," she said, "is what I call Rule #1: Time alone does not change things . . . Deliberate action by powerful administrators changes institutions." As she once told *Science*, "Changing hearts and minds one by one is much too slow. You have to change the institution, and the hearts and minds will follow."

Proof of this came when Birgeneau left MIT to become chancellor of the University of California, Berkeley, and progress subsequently stalled—the graph of newly hired women joining MIT went flat. Then a new dean arrived, more women were hired, and the line on the graph went up again.

But the MIT report also produced some unintended consequences on campuses nationwide. There was a new "Double Expectation," as McNutt calls it, that "women were expected to do everything the men did—but more of it." Women faculty were called on to mentor the growing numbers of female students, and every campus committee trying to look progressive wanted a woman on it. The handful of women faculty members available soon became overbooked and overworked. Many lost half their research time as a result of their new duties, not to mention the lucrative consulting positions their male colleagues enjoyed.

What didn't go away, McNutt said, was the "microdiscrimination—the unconscious bias that men don't know they have." The tendency to interrupt women and speak over them is perhaps the most visible sign of this. But the most disastrous is the willingness of some men to claim a woman's discovery as their own. These are only the most obvious of what is a new, twenty-first-century set of subtle biases working against women even in "light of obvious goodwill," as Lotte Bailyn, a professor at MIT's business school, put it.

Sometimes the bias wasn't even subtle. On January 14, 2005, Nancy Hopkins attended a talk by Harvard president Lawrence Summers at a closed-door conference. During his presidency, Summers had abolished Harvard's top affirmative action post and the number of Harvard's tenure offers to women had shrunk. Summers said in his talk that day that innate aptitude might explain the lack of women scientists in "the highest-ranking places," ignoring the decades of discrimination that had intentionally and unintentionally kept them out. Feeling queasy, Hopkins left the room. The media uproar over Summers's comment and Hopkins's response to it lasted for months. Many male reporters sided with Summers, and some of the men in Hopkins's department stopped speaking to her. One made threatening noises as she walked by his office. Pornographic emails poured into Hopkins's inbox.

Then one day, her old friend James Watson came by to tell her that Larry Summers was correct: girls can't do science, and she needed to apologize to Summers for suggesting otherwise. If she didn't, Watson said, he would never speak to her again.

"I lost my best friend of 40 years," Hopkins later said. That was the price you paid for making trouble for men, even if you were simply telling the truth.

As for Summers, Harvard gave him a pay raise and an honorary degree, and in 2010, he became director of President Obama's National Economic Council.

• •

The percentage of women on the faculty of MIT's School of Science went from zero in 1963 to 8 percent in 1995 to 19.2 percent in 2014. But since then, the drive toward equality has stalled. There were fourteen women faculty members in MIT's biology department in 2009—and there were still fourteen ten years later in 2019. The proportion of women faculty in biology and chemistry actually decreased during that time. At this rate, MIT estimates it will take forty-two years for women faculty members in the School of Science to reach fifty-fifty parity with men. By then, most of today's women senior scientists will be long gone.

America needs the most talented scientists and engineers from our entire population—the half that is male *and* the half that is female. We need them to address the thorny issues of global warming, safe water, sufficient food supply for the predicted ten billion people who will populate the Earth, and the wise and ethical use of artificial intelligence and powerful visualization tools for the betterment of all. These challenges will require all the talent and all the brilliance of the entire complement of humanity.

We've talked about institutional reforms to eliminate gender bias. But what about down at the level of the individual? Are women scientists—and their work—valued as highly as their male counterparts and their accomplishments?

chapter five

Cholera

I was in a small motorboat headed toward a cholera research station about a day's journey from Dhaka, the capital of Bangladesh. I was a recently minted full professor from the University of Maryland on one of my first trips to the country. As we made our way across the Ganges Delta, a speedboat-ambulance putt-putted past us, bringing a young couple and their child to the research station that also functioned as the region's cholera hospital. The parents were perhaps fifteen or sixteen years old, and visibly distraught. The mother was cuddling the baby in her arms. It was probably a boy; in 1976, many Bangladeshis did not take girl babies to a hospital.

Once the ambulance boat docked at the research station, the young parents rushed toward a canvas tent the size of a Starbucks. The tent's walls were rolled up, and I could see rows of canvas cholera cots lined up on a concrete slab floor. A funnel placed through a hole in each cot directed the patient's rice-water diarrhea into a pot on the floor. Next to each cot, a three-legged table held a basin for vomit. Measuring the liquid in the pot and the basin showed how much fluid the patient had lost and therefore how much needed to be replaced.

For cholera victims, losing seven or eight quarts of water rich with sodium chloride and potassium salts can cause electrolyte imbalance, shock, and—within hours—death. "Fine at breakfast, dead at dinner," they used to say. In 1968, less than a decade before my first visit to Bangladesh, a team of public health officers had found a cheap and easy way to treat the disease: give patients large amounts of a no-frills concoction of clean water, sugar, salt, potassium, and sodium bicarbonate (essentially baking soda). The treatment costs pennies, and its ingredients could even be mixed at home. In countries such as Bangladesh, the formula was already beginning to reduce cholera's mortality rate from more than 30 percent to less than 1 percent. With proper rehydration, the young couple's baby would almost certainly survive.

I have returned to Bangladesh almost every year since those early visits some forty-five years ago. That family energized me on a lifetime of cholera research, trying to mitigate a disease that, over the course of history and around the world, has killed *hundreds of millions* of people. My research would lead me to new theories about how contagious diseases propagate, how weather patterns and climate change can affect them, and how space satellites can predict epidemics. But persuading my fellow scientists and medical researchers of nature's role in cholera outbreaks took a quarter of a century. As I progressed in my career, I was respected as a scientific administrator, but often discounted as a woman scientist. Critics rebuffed my research publicly until, after more than two decades, it was finally accepted into the scientific canon. I believe sexism may have played a role in delaying acceptance of my laboratory's discoveries.

Here's the story as it happened.

• •

Cholera became the nineteenth century's plague, occurring along shipping and railroad trade routes around the world. Where sanitation was poor, the disease spread rapidly from person to person. But no one knew exactly *how* the disease spread.

As cholera raged through Western Europe in 1854, Filippo Pacini, a medical student in Florence, Italy, discovered its cause: *Vibrio cholerae*, a bacterium that attacks the lining of the intestines. That same year, during an epidemic of the disease in London, physician John Snow mapped the location of cholera deaths in the city and found that many victims used water from the Broad Street pump, thus establishing cholera as a waterborne disease.

In the 1870s, Edward Frankland, then the world's leading authority on contaminated water, posited that cholera occurred when sewage or manure polluted water, and invented ways to test water samples for organic nitrogen, which indicates the presence of these pollutants. And in 1883, German microbiologist Robert Koch established for certain that the disease spreads when a cholera patient's feces contaminate food or drinking water. It was not until many years later, however, in 1959, that Indian microbiologist Sambhu Nath De discovered *Vibrio cholerae*'s exact method of attack: the bacterium attaches itself to the wall of the small intestine and produces a potent toxin that makes the intestine release water and minerals so fast that the patient's volume of blood drops precipitously, impairing the blood's ability to oxygenate the body.

But cholera outbreaks can occur miles and months—even decades—apart. And not even Koch could explain where *Vibrio cholerae* went between epidemics. By the 1950s, most cholera researchers were certain—absolutely certain—that cholera bacteria survived between epidemics *only* in the intestines of healthy human beings or in contaminated food and drink, and that outside the gut, all *Vibrio cholerae* bacteria died within days. The bacterium could not live in nature. And yet this argument ran counter to a puzzling fact: epidemiologists could not find healthy humans carrying *Vibrio cholerae* between epidemics.

So the puzzle was, where did *Vibrio cholerae* hide between epidemics? Thanks to my early training in bacteriology, genetics, and oceanography—the product of my having cobbled together a path forward in a field that kept closing doors on me—I stood far enough outside the establishment to have no compunction about questioning its beliefs. And my training as a graduate student identifying marine bacteria gave me three intriguing clues about cholera's possible hiding place that most medical researchers did not possess.

The first clue involved a simple, standard test for identifying marine bacteria: whether they survive and thrive in salt water. My salt tests, crude as they were, convinced me that *Vibrio cholerae* isolated from the stool of a sick patient need salt to survive. Hence, it must be related to marine bacteria. In fact, one of my postdoctoral fellows, Fred Singleton, later showed that the cell walls of *Vibrio cholerae* actually break apart in salt-free water.

The nutrients required by vibrios (a group of waterborne bacteria) raised even stranger questions. All the vibrios I tested—including *Vibrio cholerae*—could break down chitin, the polymer that comprises the shells, carapaces, and exoskeletons of clams, oysters, mussels, crabs, shrimp, and many insects. Logically, a creature that uses chitin as food should live in a chitin-rich environment—which the human gut is not. I found myself asking, "Why would a human pathogen be digesting chitin?"

During a long day in the lab, a third intriguing clue emerged. I was conducting a standard test done on marine bacteria: turning off the lab lights and letting my eyes get used to the dark. Softly, but unmistakably, some of the *Vibrio cholerae* glowed with a ghostly light in the darkness. Many marine microorganisms are bioluminescent . . . and so were many of the *Vibrio cholerae* that cause cholera epidemics.

Those three clues—salt tolerance, chitin, and bioluminescence—were steering me toward a fundamentally different theory about *Vibrio cholerae*'s home between epidemics. They pointed unequivocally to an aquatic environment, not to the human gut.

The philosopher Thomas S. Kuhn had just written in *The Structure of Scientific Revolutions* that changing a basic scientific assumption often takes a generation. He was thinking of Max Planck, winner of the 1918 Nobel Prize in physics, who'd written with mordant humor that "a new scientific truth does not triumph by convincing its opponents and making them see the light, but rather because the opponents eventually die." In hindsight, I probably should have asked myself whether, as a woman scientist, I could hope to change the medical community's understanding of cholera within my lifetime.

• •

Shortly after I had moved from Canada to Georgetown University in 1963, I was invited to give a talk on vibrios to the Washington, DC, branch of the American Society for Microbiology. I focused in particular on one vibrio species I'd studied in Ottawa, and after my talk, John Feeley, a scientist from the NIH, asked me a life-changing question.

"You're an expert on vibrios in general," said Feeley. "Why don't you focus on *Vibrio cholerae*?" Two years before, the world's seventh cholera pandemic had started to spread from Indonesia to Western Europe and thence around the world.

"I don't have access to clinical strains," I said. My Georgetown laboratory had not been designed for experiments with potentially dangerous pathogens. It didn't have a high-end biological safety cabinet or an enclosed and ventilated workspace to protect lab workers from escaping pathogens. Nor were my students and staff trained in the highly rigorous protocol mandated for working with infectious microorganisms such as influenza.

"It's not a problem unless you drink the cultures," Feeley assured me cheerfully. "I'll send you a dozen isolates."

Sure enough, a special mailing canister soon arrived; inside were twelve test tubes containing *Vibrio cholerae* bacteria isolated from rectal swabs taken from cholera patients.

I was busy when the cultures arrived, so I put the canister in the lab refrigerator. Stored there, they would be safe until they could be transferred to freshly made growth media for study. A week later, I found time to prepare a nutritious agar-agar mixture in petri dishes. Retrieving the *Vibrio cholerae* from the fridge, I streaked the contents of each tube onto the agar and placed the petri dishes in a bacteriological incubator to keep the bacteria at 37 degrees Celsius (98.6 degrees Fahrenheit), the body temperature of a healthy human and an appropriate growth temperature for human pathogens. Then I waited the usual twenty-four hours for the vibrios to grow into colonies.

Twenty-four hours went by. Nothing happened. At forty-eight hours, there was still no growth.

I phoned Feeley. "The *Vibrio cholerae* cultures you sent me didn't survive transport."

"You put them in the refrigerator, didn't you?" He laughed. "*Vibrio cholerae* are like bananas. They don't go in the fridge. I'll send you a new set."

Like bananas? How odd, I thought. How could a salt-loving bacterium behave like a banana at low temperatures? After all, in its natural environment, a *Vibrio cholerae* bacterium would have to survive low temperatures in the winter. This was my first wisp of a clue that there might be something unusual about *Vibrio cholerae* and adverse growing conditions. It was a thread of an idea that I couldn't yet weave into the whole cloth of a hypothesis, but it was enough to make me consider joining the international hunt for *Vibrio cholerae*'s between-outbreaks hiding place.

• •

I handled Feeley's generous second shipment of *Vibrio cholerae* with great care. Bacteria have always fascinated me, and I can watch them under a microscope for hours. When I looked at the *Vibrio cholerae* Feeley had sent under my microscope, the bacteria behaved like typical vibrios, darting in one direction, then slamming on the brakes before shooting off in another. Like other vibrios, they were also wiggling constantly, their bodies contorting as their whiplike tails propelled them this way and that. (In the nineteenth century, Robert Koch had called them "comma vibrios"—"comma" for their curved shape and "vibrio" from *vibrare*, Latin for "to wave.") Testing the cultures, I discovered they had the same biochemical and physiological characteristics as the vibrios I'd isolated from fish and shellfish in Seattle and Ottawa. *Well,* I thought, *if they share the same characteristics, aren't the* Vibrio cholerae *that make people sick related to other vibrios in the environment? Marine vibrios, for example?*

I found that if I added a drop of salt-free distilled water to the cells, they burst and disappeared. The implications seemed obvious and—to a marine microbiologist—fundamentally important. All my training was telling me *Vibrio cholerae* was a marine chitin digester, recycling the components of seashells and exoskeletons. It was part of nature's carbon and nitrogen cycling, through which complex organic materials are digested down to CO_2 and N_2, which are in turn captured by and incorporated into newly developing organisms in an endless cycle. Without *Vibrio cholerae*, Earth's aquatic systems might be choked with shells from crabs, shrimp, and other creatures.

Would I be able to convince anyone, I wondered, that cholera bacteria could, and did, live in brackish (slightly salty) and salt water instead of surviving only in the human gut? To do that, I would have to show precisely where to find the cholera vibrios and explain what they do in an aquatic environment. Do they change shape? How do they survive?

And why would the enormous numbers of vibrios seen under a microscope not grow if exposed to very low temperature? These salt-tolerant, chitin-digesting, occasionally luminescent bacteria were puzzle pieces populating the back of my mind.

• •

A first step in any classic scientific investigation is to survey the published literature to see what interesting questions are still up for grabs. Before the internet, this was intensely time-consuming and tedious. However, I was in luck: the World Health Organization had recently published Robert Pollitzer's 1,019-page compendium of cholera studies conducted since AD 800. Hunting for cholera bacteria between outbreaks, physicians and scientists had searched for human carriers and environmental sources everywhere imaginable: in clean and polluted harbor water, dirt and dust, cloth and leather, rubber, paper, metals, tobacco, and every kind of food, from onions and garlic to cereals, meat, fish, fruit, wine, coffee with chicory, and coffee with milk. Searching for more scientific publications, I stumbled on a fourth fascinating clue to where cholera hid between outbreaks. The clue was in three obscure articles written by an eighty-four-year-old scientist named Frances Adelia Hallock.

I never knew what Frances Hallock looked like; I never met her or saw a photograph. She had no laboratory, she used borrowed space and borrowed equipment, and it took her fifteen years to get her research published. The great men of science would have insisted she wasn't a scientist at all. But studying those three papers, I knew what Frances Hallock was: a very smart and very unappreciated woman, one of the hidden figures of science. And although she never knew it, she helped me develop my theories about cholera, climate change, and disease transmission.

Born in 1876 in West Virginia, Hallock earned a bachelor's degree from a New England women's college, and after six discouraging years of teaching science in short-lived girls' schools, she gave up and switched to teaching Latin. At the age of thirty-one, she entered—and won—a competitive examination to teach bright working-class women at Hunter College in New York City. She wouldn't be a scientist, and she wouldn't be teaching her students to become scientists, either. There were no research jobs for women. Instead, she trained women to be technicians in public health laboratories run by men with medical degrees. Hallock remained in this job for thirty-seven years, "obsessed with the desire to make it the best course possible," she wrote toward the end of her career. "But I succeeded. I had waiting lists of several hundred names and the government ranked my course as one of the seven best in the US . . . I walked on air! My students achieved wonders after graduation, and I regarded their honors as my own."

Like any serious scientist, she chose an important topic for her students to study: the bacterium that causes cholera. Her teaching laboratory would have been quite rudimentary: a microscope, a Bunsen burner, petri dishes, test tubes, and access to an autoclave. But Hallock did something very special there. Almost all cholera researchers of the era watched vibrios under a microscope for twenty-four hours. Hallock and her students watched their vibrios daily for *six weeks*, during which time they saw the bacteria alternate "unvaryingly" between round and comma shapes. Hallock and her students made pencil drawings of what they saw through the microscope.

Others—myself included—had often seen unusually formed vibrios, especially round ones. But Hallock could be positive these sightings weren't a fluke because her students performed the same experiment meticulously year after year and always got the same results. Hallock posited the rounded form might represent a stage in the vibrio's life cycle.

Meanwhile, women's colleges nationwide were being upgraded—primarily by hiring men with advanced degrees and easing women faculty into home economics departments or early retirement. At some point, Hunter College must have told Hallock that to keep her job, she had to get a PhD; she began taking evening and Saturday classes at Barnard, Columbia University's college for women, completing her master's degree in 1919. She was promoted to assistant professor—but still wasn't considered a scientist. Taking a costly leave of absence, Hallock earned a PhD and a Phi Beta Kappa key from Johns Hopkins University. But not even an education from one of the finest institutions in the country qualified her for a job in research science. She went back to

Hunter, where she was promoted to associate professor, and continued to teach women to be lab technicians.

In 1944, at the age of sixty-eight, she retired and lived frugally at 355 East Eighty-Sixth Street in a boardinghouse with no phone line and five other lodgers who had not gone to school past eighth grade. But finally, with her teacher's pension from the City of New York, Hallock was free to become a research scientist. She continued her work on cholera vibrios, using a borrowed lab at Hunter to redo all the experiments she'd taught her students. She sketched illustrations, analyzed the data, drew conclusions, wrote them up by hand, typed them, asked knowledgeable colleagues to review the manuscripts, and then revised them. She started her first article with the words: "The concept of vibrios which has been held for 75 years requires revision." Coming from an elderly woman at Hunter College, such a statement might have seemed grandiose. She'd written only one scientific article before, based on her PhD thesis in botany; it would take fifteen years to get her cholera work published.

While Hallock was redoing her experiments, publication standards had changed: hand drawings were no longer sufficient. Instead, she needed a photomicroscope, a microscope with a camera attached, to produce acceptable images. Without one, she had no publishable evidence. But Hallock's luck turned for the better when she made friends with Dr. Clarence R. Halter, the head of radiation imaging at New York's Sloan Kettering Institute. Halter loaned part of his lab to Hallock, and Halter and Hallock took the difficult photos she needed for publication. Halter might even have recommended Hallock's articles for publication because in 1959 and 1960, *Transactions of the American Microscopical Society* published the three Hallock papers I read. Unfortunately, the journal was for people interested in microscopes, and few cholera scientists were likely to have seen Hallock's articles there. They also appeared just months too late to be included in the World Health Organization's bible of cholera research—assuming that, as the findings of a woman working outside a famous academic institution, they would have been considered for inclusion at all.

Hallock expected to be able to publish more of her research, but in February 1961, Halter retired and his laboratory was closed. "Sporting a festive carnation," Halter was given an eight-inch sterling silver bowl and a handshake at a small retirement ceremony. Hallock was left without access to a laboratory. She would die nineteen years later in a Presbyterian retirement home on Long Island, at the age of 103.

• •

I was totally enthralled by Hallock's studies. They were believable because their results were so tediously and meticulously presented over a long period of time. They also referred to *Vibrio cholerae*'s "life cycle," an idea in tune with theories I was developing around the time I discovered her work. Had Hallock's papers been published in more mainstream journals, cholera's behavior in the natural world might have been investigated twenty or thirty years earlier—and I might not have had to fight so many battles to prove my theories were right.

Curious to see the finer details of the cholera vibrio and its changes in shape, I began collaborating with George Chapman, the chair of Georgetown's microbiology department, in 1963; we published several papers together, cutting bacteria into exquisitely thin slices to observe their internal cell structure under his electron microscope. I knew the concept of stages in a bacterium's life cycle would be controversial; the "experts" often dismissed the rounded vibrios as dead or dying cells. More important, I was a woman and a junior faculty member in a Jesuit men's college in a department that had only recently begun biological research. Still, the electron micrographs Chapman and I produced did provide substantial confirmation that cholera vibrios—under specific environmental conditions that prevented them from growing and reproducing—might go into a dormant state in order to survive.

I wasn't sure yet what all I was learning about *Vibrio cholerae* meant, but I was certain of one thing. I had access to excellent research tools: computers; what may have been the best electron microscope laboratory in the country at the time; an ultracentrifuge and other advanced equipment for studying DNA; an interdisciplinary and ethnically diverse team; and the emerging computational power of DNA analysis. And in 1966, two years into my first faculty position, I received a coveted invitation to an unusual four-day workshop restricted to twenty-five marine microbiologists from around the world.

The conference was held in a hotel in Princeton, New Jersey. Professor Claude E. ZoBell of the Scripps Institution of Oceanography opened the workshop with an off-color joke about an Arabian sheikh and his harem. ZoBell, a pioneering oceanographer who'd joined Scripps in 1932, was not accustomed to having women scientists in attendance. Scripps had not allowed women oceanographers on overnight research cruises until recently, and the institution listed only one woman among ZoBell's nineteen graduate students and postdocs. I was one of two women at the conference; the other was Carol D. Litchfield, a lifelong friend who specialized in the microbiology of oceans and hyper-salty environments. A "stenotypist" named Mrs. Swanson had been hired to produce a word-for-word typescript of the conference proceedings.

NASA had organized the conference in preparation for the 1969 moon launches. Worried that the astronauts might bring dangerous microorganisms back to Earth and unhappy with the scientific advice it was getting, NASA complained that biologists worked in isolation from one another: they developed their research around one method, one technique, or one point of view, and clung to their blind spots. NASA wanted a revolution in biology, with scientists working in big, computer-savvy interdisciplinary teams. In my opinion, NASA had the right perspective.

When it was my turn to address the group, I explained that my goal was to study marine bacteria, their genetics, and their relationship to their habitat. To do that, I was merging microbiology with oceanography, genetics, mathematics, and probability. I'd even coined a term for what I was doing—polyphasic taxonomy—which I hoped would inspire microbiologists to quit quibbling over whose approach was best and instead use every tool and technique available.

Reading Mrs. Swanson's word-for-word transcript today, I'm amused by how undiplomatic an enthusiastic thirty-one-year-old can be. I questioned expert after expert, arguing for scientific precision and careful measurements: "May I ask, Dr. MacLeod, exactly what your procedure is?" and "When you say 'rapid,' you mean . . . ?"

"How do you define *Pseudomonas*—do you really mean . . . ?" I asked another researcher. When he answered, "I am only assuming that they are a *Pseudomonas* type," I warned him, "I do not think you had better say *Pseudomonas*." And so on.

Yet the conference laid the groundwork for a much-appreciated stream of funds; representatives of the National Science Foundation and the Office of Naval Research had attended the conference, and those agencies would go on to fund my laboratory for the next fifteen years. Neither agency was particularly interested in cholera, but they were keen supporters of new technology and new scientific methods. Their stable support for my lab would prove vital, because it would take years to convince my fellow scientists that *Vibrio cholerae* living wild in ponds, rivers, and marine waters could cause cholera epidemics.

But first, if I was going to determine how *Vibrio cholerae* lived between epidemics, I had to know precisely how the bacterium stayed alive in an aquatic environment throughout the year.

• •

Proving something significant in science rarely happens in a eureka moment. Scientists build incrementally, announcing each successive step in talks and publications as our research progresses. My next step was to show that the *Vibrio cholerae* isolated from cholera patients and the *Vibrio cholerae* found in nature, outside the human body, were the same species. Both a DNA study by my first graduate student, Ronald V. Citarella, and calculations done using my computer programs to determine evolutionary distances between species indicated that the two bacteria were so similar, they had to be the same species.

The stool of patients with cholera-like diarrhea frequently contained different strains of *Vibrio cholerae*: the strain known to have caused cholera outbreaks, and other strains related to those from surface water in areas of the world where cholera was common. The presence of the latter raised a question: Could vibrios from the environment actually cause cholera? Getting anyone to consider this possibility was the hard part; medical scientists like Richard A. Finkelstein already disagreed. Finkelstein, a cholera researcher at the University of Missouri School of Medicine, was one of my critics (and as noted in chapter 3 would run against me as a write-in candidate for the presidency of the ASM in 1984). He told me his PhD mentor had taught him there was only one cause of cholera and that the disease was transmitted from person to person, and he didn't plan to change his mind on either of those points.

My argument with Finkelstein illustrated the "two cultures problem" separating medical clinicians from scientists like me. Harold E. Varmus, former director of the NIH, noted once that clinicians and scientists focus on different things, with different goals and different standards for their data. Clinicians need to know as quickly as possible what's wrong with their patients and how to treat it. Scientists are supposed to test and question everything to find the bedrock truth. Vibrio specialists who'd been trained in medical schools—as most were—had been taught that *Vibrio cholerae* caused a human disease and was transferred from human to human, and that cholera had nothing to do with the environment. The differing opinions about the source of cholera would fuel decades of animosity between clinicians and scientists.

For all our differences, Finkelstein and I did have something in common: neither of us belonged to the "in" club of cholera researchers. Although Finkelstein taught in a medical school, he was not an MD, and neither was I. We were not invited, for example, to serve on the international panel that organized an important meeting on cholera. I didn't belong to the club, which was all male and medically oriented, since I was working in microbial ecology, systematics, and evolution. Finkelstein was most likely excluded because his personality was abrasive. "Finkelstein in his younger days was highly endowed with ability, but with more than his fair share of self-assuredness, and this may have abraded his well-established elders," wrote the authors of *Cholera: The American Scientific Experience, 1947–1980*. Those who knew Finkelstein personally would probably attest to that.

Sexism also worked against my attempts to challenge the theories that Finkelstein and male physician-scientists accepted as truth. During the 1960s and '70s, almost all cholera researchers were men, many of them educated in virtually female-free environments that trained their students to view women as inferiors. Sixty US medical schools still used a gynecological textbook that depicted women as either "feminine" or "domineering, demanding, masculine, aggressive, or passive." The book's 1973 edition noted that "at the core of the female personality are feminine narcissism, masochism, and passivity . . . [Women are interested in clothes, personal appearance, and beauty]." Medical research itself was so dominated by men that the NIH funded important clinical research about breast cancer, aging, heart attacks, and strokes without enrolling a single woman in their trials. Laboratory animals were more important; the NIH employed nearly forty veterinarians but only three obstetrician-gynecologists. It's not surprising that the physicians studying cholera would dismiss a new idea coming from an upstart woman scientist.

I had to choose: Should I ignore the critics or compile enough data to convince them? I had no colleagues, male or female, to back me up. If I was going to overturn one of medicine's fervently held beliefs, I'd have to do it alone.

In the end, being naturally stubborn, I decided to collect data, publish papers, and prove my hypothesis. I used each rebuff to guide my next set of experiments. Counterintuitive though it sounds, I've long thought science to be an ideal field for men and women who are underrepresented in STEMM research, whether they're white, African American, Latinx, or Native American, since science is all about fighting against the odds.

• •

I didn't know where *Vibrio cholerae* lingered between epidemics, but I was sure it was somewhere aquatic. Still, that word covers a lot of territory: oceans, bays, tidal deltas, rivers, lakes, ponds, and swamps. I knew from experiments that nearly all species of *Vibrio* respond to salt, each having its own preferred degree of salinity. *Vibrio cholerae* would likely prefer slightly salty estuaries, where tidal saltwater meets and mixes with freshwater river currents.

Happily, I was geographically well situated to explore the question. Both Georgetown University and the University of Maryland are located close to the Chesapeake Bay, the largest estuary in North America. The Chesapeake Bay is a vast system of open water and marshes fed at one end by freshwater rivers and by salty tides from the Atlantic Ocean at the other end. Baltimore, the largest city on the bay, suffered repeat cholera epidemics during the nineteenth century. And so, starting in 1963, the Chesapeake Bay became an extension of my laboratory.

Early in our studies, my technician, Betty Lovelace, and I discovered vibrios in the Chesapeake associated with oysters and with copepods, tiny crustaceans that are the size of a grain of rice but constitute the largest animal biomass on Earth. Copepods are a component of plankton—the community of tiny animals (zooplankton) and plants (phytoplankton) that drift through the upper levels of seawater, the zooplankton providing food for fish and the phytoplankton producing more oxygen than all of Earth's forests and grasslands combined.

One day in 1968, a young Japanese man appeared at the door of my Georgetown office, so tiny and unprepossessing that it had space for only a desk, chair, and bookcase. "I've come to work with Dr. Colwell," he announced politely. "May I meet Dr. Colwell?"

"I'm Dr. Colwell," I said. Judging by the look on his face, clearly he was surprised. He was a new graduate student and we had corresponded previously, but, he told me later, the idea that a woman professor was going to direct his graduate studies had never crossed his mind. Tatsuo Kaneko had come to study one of cholera's lethal cousins, *Vibrio parahaemolyticus*. This vibrio was first isolated from Japanese patients who contracted severe (and sometimes fatal) gastroenteritis after they'd eaten partially boiled juvenile sardines. I'd recently isolated the same vibrio species in water samples collected from the Chesapeake Bay, the first detection of the species outside of Japan. Today *Vibrio parahaemolyticus* is recognized as a leading cause of food poisoning from fish and shellfish in the United States.

For his PhD thesis, Kaneko collected samples of water and sediment from various parts of the Chesapeake Bay, focusing on the ecology of *Vibrio parahaemolyticus*. My research grants paid for him to join cruises on Johns Hopkins University's research vessel, the *Ridgely Warfield*, which was outfitted with advanced instrumentation and sampling gear. In the winter, Kaneko isolated vibrios from the Chesapeake's muddy bottom; during warm summer months, Kaneko isolated vibrios from plankton in the water. By demonstrating the seasonal cycle of vibrios and plankton, he provided a breakthrough strongly suggesting that outbreaks of cholera in Bangladesh might be related to the seasonal cycle of the local plankton.

Kaneko provoked an unanticipated budgetary setback in the laboratory when he also found *Vibrio parahaemolyticus* in Eastern Maryland shellfish. Crabs are culinary and economic mainstays in the state, so the discovery quickly made the news. On August 29, 1970, the *Washington Post* headlined an alarming story "Bacteria Infect Bay's Seafood." One of the grants supporting our work in those early days was $23,500 a year from a federal agency closely aligned with commercial fisheries. The morning the story appeared, even before I could pick up the newspaper from our driveway, an agency representative telephoned. The grant was canceled—the publicity was bad for local fisheries. The sum involved was relatively small, but I was stunned that the work was being viewed so politically. But all turned out well: When I reported on Kaneko's research a short time later at a major scientific meeting, a representative of the National Oceanic and Atmospheric Administration (NOAA) suggested I apply there for support. I did, and the NOAA awarded me a new grant worth $250,000, roughly equal to $1.6 million today.

An idea was beginning to form in my mind: Did *Vibrio cholerae* go into some sort of wintertime hibernation, hiding in muddy river bottoms until more congenial weather returned? It wasn't yet a testable hypothesis, much less a complete theory. But it was another thread—like refrigerated bananas, salt tests, chitin digesters, and Frances Hallock's round shapes—that might one day be woven into something larger. Progress was being made.

• •

Within a few years of moving to the University of Maryland in 1972, I had what was, in its day, a big lab, filled with twenty-five to forty graduate students, postdoctoral fellows, and visiting scientists. Joining a major research university with a warm, supportive community of scholars had given an extraordinary (and interdisciplinary) boost to my research. And being surrounded by colleagues with similar interests as part of a microbiology department (rather than a general biology department) was truly catalytic. Soon I was working on oil degradation, effects of deep sea pressure on bacteria, mercury resistance and metabolism, and, of course, copepods and vibrios.

In science, we almost always work collaboratively, often shoulder-to-shoulder, days and nights. For a while, until our lab space was expanded, some students worked in shifts, cleaning their bench at the end of the day for another student to occupy overnight. I no longer had time to give all the hands-on instruction that students needed, so Janie Robinson, the tech who'd moved with me from Georgetown, taught newcomers the rules of the lab. Graduate students learned basic microbiological techniques from postdoctoral fellows, and a senior postdoc managed the lab day to day. New students could talk to team members before choosing their own research projects. You can dream in science, and if a new student had a burning interest in a subject no one was already studying, I'd write a grant application for funding. I loved working with the students, making sure that appropriate methods and tools were used, helping to analyze and interpret data, and relating the results to a bigger picture. When more advanced technology was developed, whether it was a novel spectrophotometer, centrifuge, or gas chromatograph, I'd find a way to get it so we could produce more precise data or tackle a new problem. With more than $2 million in federal grants (approximately $6 million today), I could pay for students' stipends, travel expenses, time on research ships, and the latest equipment. Serving on national and international panels, I'd often find researchers willing to collaborate—Japanese scientists such as Kazuhiro Kogure at the Ocean Research Institute, University of Tokyo; Carla Pruzzo of the University of Genoa, Italy, who made new discoveries on the effect of climate change on pathogenic vibrios; and French colleagues in Marseille (Micheline Bianchi), Brest (Monique Pommepuis and Dominique Hervio-Heath), and Montpellier (Patrick Monfort), who partnered on the discoveries of vibrios in the environment. Over the years, I've had the good fortune to serve as advisor for many excellent PhD students, all of whom have found positions, whether in academia, government laboratories, industry, or investing—even in winemaking and art. Four have been elected to the National Academy of Sciences, including Jody Deming, who is now professor of oceanography at the University of Washington. Jody and a fellow graduate student, Paul Tabor, did extraordinary research on pressure-loving bacteria isolated from the deepest regions of the ocean.

In short, by the mid-1970s, I had good reason to believe I had climbed sufficiently high up the academic ladder to escape the belittling men in my field. It was a busy but glorious time. Sure, I wasn't in a hot specialty, but this meant that without lots of competitors, I was free to forge my own path. I was coming to realize that vibrios truly are wild, environmental bacteria—a revelatory way of seeing the world. I was happy with the research my students and I were doing, contributing to scientific knowledge relevant to human health.

• •

Early in 1975, a long-haired young man in shorts and flip-flops knocked on my office door and announced that he wanted to enroll in graduate school and work in my laboratory. James B. Kaper was working as a carpenter and stage technician for a local opera theater, and his résumé showed that his undergraduate grades were not the highest. When he took my microbial systematics course, however, he studied hard and got the highest grade in the class, so I knew he was bright. He was also determined.

Kaper took Tatsuo Kaneko's work several steps further. Kaneko had studied only a few places in the Chesapeake, but Kaper studied the length of the bay. And he focused on cholera. He found what he believed were *Vibrio cholerae* in sixty-five samples of water, sediment, and shellfish collected from the Chesapeake Bay. Back in our laboratory, he positively identified cholera vibrios in the samples. He found the largest number of cholera vibrios mid-bay, where moderately salty water from the Atlantic mixes with freshwater from rivers. In summer, he found *Vibrio cholerae* on plankton copepods; in winter, he discovered them attached to copepods burrowing into the muddy Chesapeake sediment. And in a crucial piece of evidence against the prevailing theories of how cholera spreads, Kaper found most of the *Vibrio cholerae* in water without fecal contamination, suggesting that the vibrios had not entered the Chesapeake in sewage or cholera patients' stool.

As I noted earlier, the Baltimore area had suffered numerous cholera outbreaks during the nineteenth century, but we still could not say positively that environmental *Vibrio cholerae* could cause epidemics. DNA analysis suggested that the environmental *Vibrio cholerae* we isolated from the Bay could cause cholera, but an expert told me he couldn't be certain the isolates from the samples we'd sent and he'd tested had produced cholera toxin. (The expert later admitted he'd withheld confirmation because he didn't want to get involved in any controversy.) As a result, the article we published in the October 28, 1977, issue of *Science* could announce only that Kaper had found *Vibrio cholerae* in the Chesapeake and that the discovery raised "many questions, some very disturbing." In a wishy-washy compromise, we had to add that these vibrios were "not recognized as a serious epidemic threat, although they have caused cholera-like diarrhea sporadically."

Kaper's discovery of *Vibrio cholerae* in the Chesapeake sharply divided the cholera research world. We worked in a kind of purgatory, presenting papers, speaking at meetings, and publishing—sometimes with difficulty, and sometimes not. Many vibrio scientists agreed with me that, of course, *Vibrio cholerae* are naturally present in estuaries around the world, from the Bay of Bengal to the Chesapeake Bay. But the reaction of some in the medical fraternity was vociferously negative. Cholera is a human disease, they insisted. *Vibrio cholerae* in the Chesapeake, when there's no cholera epidemic there? Ridiculous! The field of cholera microbiology was so divided that John G. Holt, the longtime editor of *Bergey's Manual of Systematic Bacteriology*, the bible of bacterial taxonomy, said he couldn't find a neutral expert to write about vibrios for a new edition of the book. I was much too controversial to be asked.

• •

I was an easy target for jokes. When a meeting in Quebec City included a recreational tour, I was asked, as one of the organizers, to speed things up. "Guys," I said, "let's get on the bus." Quick as a wink, a particularly august Brit said, "Oh, we've got a dolly bird to lead us," or words to that effect. Kaper remembers a well-known speaker from the CDC getting drunk and loud at a major meeting in Las Vegas. "I can speak over the noise," the speaker bragged, slurring his words, "but Rita's just a little girl." It's hard enough to forget bullying when you hear it yourself, but when your students hear it, too? It's incredibly demoralizing.

Some graduate students, and not just those in my laboratory, were teased for having a female thesis advisor. When one of my students gave a talk, the author of a popular microbial ecology textbook stage-whispered disparaging comments all through my student's presentation—just like the arrogant Berkeley professor who spoke over my talk many years before. Richard Finkelstein was still a frequent critic. One of my students still tells the story of how Finkelstein told him, in public, that if he wanted a PhD, he should change the subject of his dissertation. And Anwar Huq, my longtime collaborator and future lab manager, remembers my speaking at an NIH conference and Finkelstein interrupting to say, "Oh my God. The way you're talking, it looks like vibrios are even in your backyard."

"I haven't checked there yet," I shot back.

While my colleagues' criticism was hard to take, it was often useful, pointing me to the next question I needed to answer. At one conference, after Huq gave a talk, he was asked, "How do you know that *Vibrio cholerae* from Bangladesh behave exactly like those in the Chesapeake?" I was listening from a seat at the back of the room, so I stood up and said, "Anwar, Dick is right. We'll go to Bangladesh and look for that." So we did. Huq collected the evidence that *Vibrio cholerae* in both the Bay of Bengal and the Chesapeake Bay are associated with copepods, and that the annual cycles of plankton, including copepods, coincide with seasonal peaks in *Vibrio cholerae* abundance in the water . . . and that the peaks coincide with cholera outbreaks in the Bangladeshi villages where we worked. *Vibrio cholerae* really was a global aquatic pathogen.

To this day, I do not believe that I—or my students—would have been treated with the same disdain had I been a man. When people ask me how I had time to publish more than eight hundred scientific papers over the course of my career, I tell them I had no choice: as a woman, I had to prove my findings twenty times over just to get them taken seriously. Proving, proving, proving—you were always swimming against the current. It does wear on a person after a while. I'd come home from confrontational conferences feeling angry and resentful, and my husband, Jack, would say, "Just ignore them. You'll feel better if you do." Other times, he'd listen as I raged, and then point out the absurdities and pettiness in my colleagues' attacks, suggesting logical rebuttals. Then we'd go sailing for a weekend, and with the children and a strong wind and full sails, the world was a happy place again.

Jack and I were racing our seventeen-foot sailboat in regattas almost every weekend by then. Jack was the strategist and I was his crew, hanging on, often drenched to the skin, flying the spinnaker and handling the jib during sharp tacks to outmaneuver our competitors. During one race in Massachusetts, we sailed through a terrible storm—the rain was pouring down and the wind was blowing at 30 knots, and Jack kept shouting, "Keep the spinnaker flying, keep it trimmed." To keep the spinnaker full, you had to keep your eye on the sail and adjust according to the wind. When we'd rounded the mark and the spinnaker could be dropped, I looked back, and saw that half the boats in the race had capsized. "That's why I didn't want you to look back," he said with a big grin on his face. Jack was always surprised that I never really felt secure hanging out over the water with my feet planted on the rail. There I was: a marine microbiologist who, when the wind reached 25 knots, would white-knuckle through every wild tack. But I persisted. He knew I really liked to win the regatta—and to be with him.

• •

I'd taken my first working field trip to Bangladesh back in 1976 determined to study *Vibrio cholerae* in its natural environment and obtain new isolates to bring back to the lab. I knew it would be challenging. The water in Bangladesh is in a constant state of flux: at various times of the year, Himalayan snowmelt races down through the severely polluted rivers of Nepal and India; salty ocean tides rush in from the Bay of Bengal; heavy monsoon rains flood the land and churn up the muddy bottoms of rivers, ponds, and wetlands; and temperatures and water levels rise and fall with currents, tides, and seasons. And my field laboratory was rudimentary: a microscope, a small incubator, an alcohol lamp to sterilize inoculating needles, and a pressure cooker to sterilize media.

On a visit to Bangladesh some years after that first field trip, I learned about some extraordinary cows from William "Buck" Greenough, a Johns Hopkins medical school physician who helped develop the life-saving oral rehydration treatment for cholera patients that I mentioned at the beginning of this chapter. Greenough told me that in 1963, these remarkable cows were discovered to have antibodies to a disease that cows never get: cholera. The antibodies meant that somewhere, the cows must have met up with *Vibrio cholerae*. I was sure the cows had come in contact with vibrios while grazing along the side of an estuary and drinking the brackish water. "Let's investigate," I suggested. Years later, Greenough still remembered pulling on boots and going to collect water samples with me.

Back in the field laboratory, looking through a microscope constructed for viewing fluorescent objects, Greenough and I could see brightly glowing rod-shaped vibrios. There weren't many of them, but they were there. My Maryland team had developed a cholera antibody in rabbits and then chemically linked it to a molecule that would glow under an ultraviolet light microscope. (Antibodies are proteins produced in the body that react with specific foreign agents, like bacterial cells. When linked to a chemical compound that lights up under ultraviolet light, antibodies allow researchers to see the bacteria in raw water samples under a UV microscope.) With the fluorescent antibody coating the vibrios, they looked like glowing commas. We could actually see *Vibrio cholerae* that inhabited an estuary in Bangladesh. Years later, Greenough recalled our excitement. Critics said it was impossible to detect *Vibrio cholerae* in water samples from the environment—but we'd done it. Greenough predicted trouble convincing the skeptics.

Back home, when I mounted the photographs on slides and showed them in amphitheaters using 1980s audiovisual equipment, they looked like green snowflakes against a dark background. Any skeptic could say, "They're not really *Vibrio cholerae.*" If our critics could have isolated wild *Vibrio cholerae* from the environment themselves and grown the bacteria in a test tube or petri dish in their own laboratories, they would have been convinced. But the orthodox view among scientists was, "If I can't grow a bacterium in the laboratory, it is not living. It is dead." Microbiologists knew how to grow *Vibrio cholerae* from patients' stool samples or rectal swabs, but they didn't understand enough about how the bacteria exist in the wild to be able to bring them from their natural habitat and get them to grow in a lab.

Reproducibility is the key to success in science. When people share data and reproduce the key observations in one another's experiments, we can be pretty sure we're right. Our laboratory hid nothing: we published all our results, and we had visitors all the time. The lab was a veritable United Nations, with scientists from all over the world. But no one ever visited from Woods Hole or Scripps Institution of Oceanography or even from the University of Washington, where I'd started my research. Without researchers from leading US institutions collaborating in our experiments, it was difficult to convince many microbiologists of our findings.

We also still needed to answer the big question about cold weather. Why could we find large numbers of cholera bacteria attached to their copepod hosts in water samples from the Chesapeake Bay and the Bay of Bengal in warm weather but not in cold-water samples? Skeptics said wintertime cholera bacteria were sick, dying, or dead. We knew, though, that their cells remained intact and attached to copepods that were "overwintering" in sediment. Besides, if the vibrios were dead, how could they come back in such large numbers in the summer? It was frustrating: we could detect the bacteria in samples of sediment collected during the winter, but using the regular methods, we couldn't get the bacteria to grow in the lab.

By now, I was hypothesizing that a bacterium could look dead and refuse to grow in a lab, but still be alive and capable of causing disease. The idea went against everything we'd been taught. The accepted dogma said that only bacteria with a spore stage in their life cycle could go dormant; I'd been taught that as an undergrad. That meant that vibrios, which don't produce spores, could not go dormant. I'd also been taught that pathogens that can't be grown in a laboratory were either dead or dying.

But I was beginning to think the cholera bacteria must be hunkering down in the sediment with their copepod hosts as a survival mechanism in wintertime. They looked dead: round, shriveled, and small. And when we brought them into the lab and fed them one-size-fits-all lab food, they wouldn't grow or reproduce. If conditions were more congenial, I wondered, would they resuscitate, grow, reproduce, and again become potentially lethal? I realized my hypothesis was so contrary to accepted wisdom that it would polarize scientific opinion, and knew I'd have to proceed carefully, documenting each step of the way.

• •

We still didn't know what precisely would make a cholera vibrio go dormant. But before we could figure that out, we needed to develop a test that would let us—and our critics—see cholera bacteria in their dormant stage in the wild. It took graduate students and postdocs from my lab five years, but they succeeded. And a visiting scientist from China, Huai Shu Xu, improved the method for field use.

The first important test of our new method came when Nell Roberts telephoned and introduced herself as a laboratory technician and instructor in Louisiana's public health department. "Dr. Colwell, I've been following your work for years," she said. "I'd like you to check out a bayou for cholera vibrios. We have some cholera cases here in Louisiana, and I think I know where they came from."

The United States had had almost no cholera cases since 1914, but now, suddenly, eleven crabbers who'd picnicked along the Gulf of Mexico had wound up hospitalized with what was clinically confirmed to be cholera. Roberts knew the crabbers had picnicked near a bayou in Lake Charles, Louisiana, and she was sure disease-causing *Vibrio cholerae* were naturally present in brackish water there. She wanted me to come to Louisiana and find them. I told her that Xu and I would fly down.

The three of us drove to the bayou, collected samples of the water, and, back in Roberts's laboratory, put droplets of the sample on slides. Next, we added our specially prepared fluorescent antibody that would attach only to *Vibrio cholerae*. We adjusted the field of vision in Roberts's epifluorescent microscope. And then we saw them: the unmistakable outlines of Koch's comma-shaped *Vibrio cholerae* cells. Coated with the fluorescent antibody, they looked like bright, twinkling little green stars. Critics had said any *Vibrio cholerae* found in natural waters must be dead or dying, but these vibrios were clearly intact and alive. On some of them, we could even see flagella, the threadlike appendages that propel vibrios through water. Thrilled to see disease-causing cholera bacteria in water from an American bayou, we whooped and hollered and waltzed around the lab. "I told you it was there!" Nell Roberts kept saying. "I knew it was there!"

It wasn't the greatest scientific discovery in the world, yet I can only describe the moment as transcendent, as if we were peeking inside a tiny part of nature to glimpse the engine driving it all. We had strong evidence that *Vibrio cholerae* from the bayous had quite probably made those crabbers sick. And soon, other scientists were confirming our findings by discovering *Vibrio cholerae* in brackish waters around the world.

The paper we wrote about those Louisiana crabbers in 1982 should have ended ten years of controversy. But it didn't. To change minds, I knew we had to get our research published in a peer-reviewed journal. It didn't have to be the most famous one—we just needed it published *somewhere*. After that, we could write a review of the entire field for a major journal and place our work in context. I submitted the paper to a colleague who was guest-editing an issue of a fairly new journal, *Microbial Ecology.* That colleague, Samuel W. Joseph, understood our work and noted in an editorial, "Increasingly, we are finding that emerging diseases of man are caused by organisms ubiquitous in the environment." Ecology and medicine were finally moving closer together.

• •

We'd shown that disease-causing *Vibrio cholerae* could be isolated from the environment. But we still couldn't demonstrate a key part of my hypothesis: namely, that under certain conditions, environmental *Vibrio cholerae* go dormant and then revive. Darlene Roszak, who was doing her PhD with me, agreed to work on the problem at the molecular level. Roszak had earned her bachelor's degree as a thirtysomething widow with four children. She remarried, and her new husband, Mike MacDonell, a Vietnam veteran, who was very proud and supportive of Darlene, was also working on a PhD with me.

Roszak started by keeping environmental *Vibrio cholerae* at cold temperatures until they stopped growing. Next, she exposed them to amino acids that had been tagged with a radioactive isotope, making it easy to monitor how much radioactive CO_2 was released by the vibrios. Then she added nalidixic acid, an antibiotic that prevents cell division but not metabolism, to the bacteria and observed the effects under a microscope. She watched as the vibrios grew longer and longer—to do that, the cells *had* to be alive. The pièce de résistance came when Roszak combined the experiments, tagging the cells with radiolabeled amino acid and exposing them to nalidixic acid, before placing them on photo emulsion film. As the bacteria metabolized, growing larger and longer and releasing radioactive CO_2, they essentially photographed themselves—like selfies in science.

Roszak's experiments showed that once the cells were exposed to cold, they would not reproduce in the laboratory, *but they were not dead*. They were in a kind of stasis, like an idling car engine: the car's not moving, but it's not dead, either—the engine's just turning over and over, keeping the machinery going. In the case of *Vibrio cholerae*, the bacteria idled until new nutrients arrived and the cells could replicate again. We called the phenomenon "viable but not culturable," or VBNC for short. After the usual heated controversy, Darlene Roszak's paper on the subject became a classic. Microbiologists accepted that there are far more bacteria in nature than they'd thought. All those years, they'd been studying 1 percent or less of the bacteria in the environment: the 1 percent we call "lab weeds" because they grow so easily in the lab.

A cascade of studies from our laboratory and others demonstrated that our VBNC hypothesis was correct. In bad times, *Vibrio cholerae* and many other bacteria that do not form spores go dormant in brackish waters. In good times, they revive and, if ingested, can cause disease. In 1985, Charles Somerville and Ivor T. Knight, then students in my lab, used the recently published polymerase chain reaction method to show that environmental *Vibrio cholerae* can carry the toxin-producing gene associated with epidemic-causing strains of the bacterium.

Mike Levine, an outstanding scientist and medical doctor at the University of Maryland School of Medicine who is also a good friend, agreed to work with me to carry out a clinical trial with volunteers who consented to ingest small amounts of dormant cholera cells of a less virulent strain. One volunteer got mild diarrhea, and others passed the weakened but epidemic-causing cholera cells in their stool. Other investigators have since shown that more than fifty disease-causing bacterial species—including *Mycobacterium tuberculosis*, *E. coli*, and *Helicobacter pylori*, as well as species of the genera *Legionella*, *Salmonella*, *Shigella*, *Campylobacter*, and *Chlamydia*—enter a VBNC state when stressed by food insecurity, antibiotics, very high or very low temperatures, desiccation, carbon or nitrogen starvation, heavy metals, white light irradiation, or UV light. And with the advent of metagenomics—the study of all the microorganisms present in samples collected directly from the environment—everyone now looks for "unculturables," species that are difficult or impossible to grow in the lab.

In 1996, while doing research for a speech I was writing, I sat down to count the number of microbiology papers focused on the VBNC theory of dormancy. It had been more than a decade since I'd coined the term "VBNC" and a quarter century since two wonderfully supportive men, my PhD advisor John Liston and Canadian microbiologist Norman Gibbons, had helped me circumvent anti-nepotism rules to continue my work on vibrios. To my surprise, I counted several hundred papers on VBNC. Despite the medical community's animosity, the paradigm had been shifting under my feet. There'd been no big eureka moment. Each experiment had changed a few minds, until the accepted wisdom had slowly been transformed. By 2015, more than six hundred studies of VBNC had been published, and since then, other researchers have named variations of the VBNC phenomenon "persisters," "unculturables," and "viable but *not yet* culturable."

The scientific revolution had taken just about as long as Thomas Kuhn hypothesized it would.

• •

In my view, cholera is a vector-borne disease—like malaria, spread by mosquitoes, or Lyme disease, carried by ticks. The vector, in the case of cholera, is the copepod, those tiny crustacean zooplankton. A single copepod can carry fifty thousand *Vibrio cholerae* bacterial cells. If you ingest untreated water with copepods in it, the acid in your stomach will digest the copepods but will not kill the bacteria they carry, leaving the *Vibrio cholerae* free to attach themselves to the lining of your small intestine (as if you were a giant copepod).

Vaccines can be used to eradicate smallpox and polio because between epidemics, those pathogens survive primarily inside people or animals, but we can never eradicate cholera because vibrios are part of aquatic ecosystems around the world. How, then, I wondered, could we put our scientific research to practical use? We may have changed a century-old medical dogma, but we still had not helped any of the thousands of people suffering from cholera around the world.

The people who suffer most in a cholera epidemic are women and children. To understand why, imagine standing at the edge of a small pond in a remote area of Bangladesh. In a far corner of the pond, a latrine is draining into the water. Nearby is a cowherd, steering his cows in for a drink and a wash. A woman kneels at the water's edge, washing dishes. Beside her, a little girl collects drinking water for her family. Throughout the developing world, hauling water is women's work. Women and school-aged girls, pulled out of their classrooms, fetch water for their families, cook food, care for sick relatives, wash soiled bedding and clothing, and dispose of diarrhea and vomit from cholera patients. The same is true in understaffed rural hospitals. Richard Cash, a Harvard Public Health researcher who helped develop the oral rehydration formula in the 1970s, had shown that ingesting just a teaspoon of water contaminated with *Vibrio cholerae* bacteria could cause cholera—that single teaspoon could contain a million or more cholera bacteria.

Rich countries—including the United States, numerous countries in Europe, and Japan and Singapore—control cholera by filtering, chlorinating, and safely distributing water. This process removes pathogens attached to particulates and inactivates or kills free-swimming microorganisms not trapped by filtration. With this treatment, at least twenty-seven waterborne diarrheal disease agents are removed from the water supply. In an attempt to provide similarly safe drinking water to Bangladesh during the 1960s, the World Bank funded projects to drill extremely deep wells down past polluted surface waters into what was assumed to be unpolluted groundwater. Tragically, however, Bangladesh has the most arsenic-rich groundwater in the world, due to natural sources of the element in the country's soil. Drinking water that contains high levels of arsenic and eating food, such as rice, cooked in water have given Bangladeshis high rates of arsenic-induced cancer. Among the side effects of arsenic consumption are hair and tooth loss.

In the late 1980s, my colleague Anwar Huq and I decided to see if we could use our science to help provide remote villages with safer surface water to drink. In his childhood in Bangladesh, Huq remembered seeing village women using some kind of filter for their drinking water.

With colleagues at the International Centre for Diarrhoeal Disease

Research in Dhaka, Bangladesh, Huq and I began experimenting with various kinds of filtering material. Nylon strainers were used in Africa to remove the cyclopes that carry Guinea worm larvae; under the skin, the larvae can grow into parasitic worms two or three feet long before they emerge, generally through the feet. But nylon mesh fabric was too expensive for most Bangladeshi villagers, and the material of T-shirts worn by many men in the region didn't work well as a filter.

Finally, we discovered that folding cotton sari cloth four or eight times creates a mesh fine enough to trap particulate matter, including copepods laden with *Vibrio cholerae*. Almost every Bangladeshi household has old cloth on hand since sari cloth is the material of choice for many women in Bangladesh and India. Further tests showed that when sari sieves were used, the number of cholera cases could be halved. But we were told it would be culturally unacceptable for men to drink water filtered through a sari sieve because used sari cloth was considered unclean. We also had trouble obtaining funding for such a low-tech project that didn't involve a marketable device or instrument. However, with financing from the Thrasher Foundation and the National Institute of Nursing Research at the NIH, we trained Bangladeshi women how to use sari cloth filters. These women then acted as "extension agents," teaching women in other villages the filtering technique and following up with weekly visits to encourage compliance. In the sixty-five villages we studied, sari cloth filters nearly halved the rate of cholera. Five years later, we published the results of a follow-up study of 7,122 village women showing that the method was sustainable: nearly three-quarters of the women still filtered their families' water through sari cloth.

• •

In January 2010, Haiti was struck by a devastating earthquake measuring 7.6 on the Richter scale; more than fifty aftershocks followed. An estimated 200,000 people were killed, 300,000 were injured, and one million were rendered homeless, and Haiti's already grossly inadequate sanitation and water treatment infrastructure was heavily damaged. The summer following the earthquake was Haiti's warmest in fifty years, and in November of the same year, Hurricane Tomas delivered the country's heaviest rainfall in half a century, causing massive flooding. Refugees crowded together in camps without adequate safe water or sanitation. It was an ecological prescription—a "perfect storm"—for a cholera epidemic to strike. And strike it did, sickening more than 800,000 people and killing nearly 10,000.

What caused the outbreak of cholera? Had indigenous *Vibrio cholerae* played any role? During the first weeks of the epidemic, we arranged for stool samples to be collected from eighty-one cholera patients in eighteen towns along the Haitian coast. Half of the patients were victims of cholera strains known to have caused epidemics in Southeast Asia and Africa. But one in five patients were not. Ten coauthors and I wrote a paper about the possible role of native, environmental bacteria in exacerbating Haiti's epidemic. It was impossible to prove that cholera vibrios inhabited Haiti's coastal waters or rivers because shipping samples out of Haiti's airport or seaport was banned shortly after the epidemic began, and our paper received a storm of criticism.

Two years later, nutritionists from Cornell and the University of Virginia revisited a study of 117 HIV-exposed infants in urban Haitian hospitals. As part of the study, the scientists had frozen and archived 301 samples of the children's stool. Reanalyzing the samples after the initial study, the scientists found *Vibrio cholerae* in the stool of nine of the infants, meaning the bacterium was present two years *before* the earthquake.

Since the Haitian earthquake, we've had to revise many hypotheses we thought were sacrosanct. For instance:

- There isn't just one kind of cholera epidemic. The one I'd studied for years in Bangladesh appears near coastlines every spring and fall when plankton populations bloom and tides drive seawater inland. Anwar Huq and I teamed up with Antarpreet S. Jutla, a brilliant young hydrology engineer from India, and Elizabeth Whitcomb, a physician-historian from New Zealand, to analyze some nineteenth-century British Army maps showing the location of each and every cholera death in India from 1876 to 1900. Whitcomb is a meticulous scientist and a brilliant diagnostician, and the data revealed that massive epidemics occur near inland tidal rivers episodically. These brief but savage epidemics in the country's interior tended to occur when large numbers of people crowded together—as they did during religious festivals, wartime, or natural disasters—and, as we found by cross-referencing this information with British meteorological data, when extremely hot weather was followed by heavy rains, flooding, a lack of safe water, and sanitation breakdowns. Authorities are unprepared for such inland epidemics, so mortality rates can be very high. Haiti's tragic epidemic followed a similar trajectory: an earthquake that forced many people into crowded refugee camps, followed by a heat wave and heavy rain.

- Our definition of "cholera" may also need to be revised. We now know that a patient with watery stools may not be suffering from cholera alone. The patient may have a mix of infections caused by *Vibrio cholerae*, *Shigella, E. coli*, and *Salmonella* bacteria, viruses, fungi, and/or parasites, what's known as a polymicrobial infection.
- *Vibrio cholerae* can transfer approximately 80 percent of their genes to and from nearby cells and even nearby species (like *E. coli*). Such genetic fluidity explains how cholera vibrios can survive the radical changes in their environment, including daily tides and seasonal changes in water salinity, nutrients, depth, and temperature, as well as air temperature. It also explains how some *Vibrio cholerae* cause disease, while others do not, and how virulent strains of cholera have recently emerged.
- Today, three out of four countries with cholera cases are in sub-Saharan Africa, where fatality rates are triple those in Asia. Haiti, according to the World Health Organization, had no new cholera cases reported in 2019.
- With global warming, vibrios and vibrio-caused diseases—including cholera—are spreading north in

> the North Atlantic and the Baltic Sea regions. People who've gone wading in the Baltic during heat waves have been hospitalized with cholera and other vibrio infections, and several have died.

• •

I've spent a long time thinking about how to predict cholera epidemics. Countries with limited budgets need all the advance warning they can get to mobilize physicians, nurses, public health workers, oral rehydration kits, antibiotics, water purification equipment, vaccines for children and the elderly, hygiene kits, and public education programs.

When NASA's Landsat satellites started collecting data on Earth's natural resources in 1972, they spotted—*from space*—huge green mats of plankton floating along tropical coastlines. The satellites couldn't measure animal life in the mats, much less the number of copepods hosting *Vibrio cholerae*. But they *could* measure chlorophyll pigments and the temperature and height of the sea surface. I'd been studying the connections between cholera vibrios, rainfall, salinity, available nutrients, and air and water temperatures since the 1970s. And I wanted to use NASA's satellite data to create a timeline to predict the increase in chlorophyll far enough in advance to issue cholera alerts. Plankton scientists already knew that sunlight intensity varies with the seasons. When a region experiences more daylight hours, the surface water warms, and phytoplankton (the "grass of the sea") explode in growth; copepods and other microscopic animals that comprise the zooplankton feed on phytoplankton and reproduce, and roughly four to six weeks later, zooplankton populations will have peaked and then crash. Shortly thereafter, villagers who rely on untreated pond water would be drinking water containing the vibrios that were released.

The world's climate is changing, and our oceans are warming. The need for a cholera prediction system is even greater than before. Warmer ocean water will affect plankton populations and may possibly result in larger populations of vibrio bacteria and perhaps more and longer cholera epidemics in the developing world. During the 1980s, scientists estimated that by 2050, rising sea levels would inundate 17 percent of

Bangladesh's land, displacing 18 million people and causing vast migrations that could potentially destabilize other countries—all within the lifetime of my children and grandchildren.

In 1982, I began a very productive collaboration with NASA scientists Byron Wood, Brad Lobitz, and Louisa Beck. Their technology was not yet ready to predict an epidemic. Satellite data could not yet be downloaded to a computer in a usable format for easy analysis; it was still transported on large reels of tape and disks. By the late 1990s, however, the data were digitized, and I worked on computer models with the team at NASA's Ames Research Center at Moffett Federal Airfield in California. Subsequently, Guillaume Constantin de Magny, a French postdoctoral fellow, and other students in my lab used pre-epidemic rainfall, air temperature, water salinity, and river depth to improve computer models for predicting cholera epidemics. The depth of rivers is surprisingly important because as water levels drop, rivers become saltier and more turbulent, churning up bacteria-rich sediment from the river bottoms. By 2008, our models could predict an increase in the number of cholera cases with uncanny accuracy. NASA's technology kept pace, and by 2011, its environmental satellites were so sophisticated, they could determine the surface temperature of a square kilometer of ocean anywhere on Earth, every single day.

Four years later, my colleagues and I assembled a computer model to predict where and when a cholera epidemic was most likely to occur by using Antarpreet S. Jutla's mathematical skills, Anwar Huq's knowledge of cholera and Bangladesh, and my understanding of microbiology, microbial ecology, and molecular biology. Our first predictions were for the Chesapeake Bay, Zimbabwe, Mozambique, and Senegal. Then, in 2017, the biggest cholera outbreak in recorded history struck tiny Yemen, one of the poorest countries in the world.

Yemen sits at the southern tip of the Arabian Peninsula. The country is home to four UNESCO World Heritage treasures, but civil war and bombing by the US-supported Saudi Arabian Air Force had destroyed much of the country's sanitation and water treatment infrastructure. Cholera followed, in what the United Nations and the World Health Organization called the worst humanitarian crisis on the globe. More than 1.2 million of Yemen's 26 million people were diagnosed with cholera, a third of them small children. To date, more than 2,300 people have died from what is a preventable and cheaply treatable disease.

Dividing Yemen into regions the size of a typical US county, Jutla's increasingly sophisticated model predicted—with 92 percent accuracy—the cholera risk for each region. When Fergus McBean, a Scottish humanitarian advisor with the British Department for International Development, read about our work in the late fall of 2017, he issued us a challenge: get a cholera-forecasting system for Yemen up and working before the start of the next rainy season. That gave us four months.

It was working in three.

In March 2018, one month ahead of the rainy season, the Department for International Development began using our model's forecasts. Early results showed that our predictions, coupled with weather forecasts from the UK Meteorology Office, were helping UNICEF and other aid groups target their response where support was needed most. Dressed in a handsome tartan kilt and sporran, McBean received an award from the Department for International Development for his role in controlling the outbreak. The work, he said, "could not have been completed without a team effort" involving the University of Maryland, the University of West Virginia, NASA, and UNICEF.

Modern science has become truly interdisciplinary. Solving the complex problems of the twenty-first century will require integrating the traditional physical, chemical, and biological sciences with the social and behavioral sciences. It will take collaboration across fields of research and dialogue between and among diverse groups of people.

So let's look next at what women scientists can do if they control the money and have a sisterhood to help.

chapter six

More Women = Better Science

On November 27, 1998, not long after I had been sworn in as director of the National Science Foundation (NSF), I arrived in my office to learn that Vice Admiral Paul G. Gaffney, the US Navy's chief of research, was on my calendar for an 11:01 a.m. to 12:01 p.m. appointment.

After the usual social niceties, Admiral Gaffney, a personal friend, told me that the NSF was funding a team of University of Hawaii oceanographers to spend two weeks aboard a US nuclear submarine to map the Arctic seafloor. But there was a problem, the admiral explained. The scientific team had chosen an excellent oceanographer, Associate Professor Margo H. Edwards, as its chief scientist. But women were not allowed on US submarines.

"Well," I said, "no woman, no money."

The American oceanographic community and the US Navy had banned women scientists from scientific cruises until the late 1960s, when occasional women began to be allowed on board. But change was gradual. The "US Navy [w]as a much more impenetrable barrier for women than even industry or academia," oceanographer Kathleen Crane recalled in her memoir. Crane worked in the Arctic for more than twenty years, using her grants to pay for berths on Russian, Swedish, Norwegian, Canadian, French, and German ships. Cleaning women could work in American submarines while they were in port, and female contractors could even stay overnight to check equipment.

But a woman scientist? On board a nuclear submarine during a naval operation? No way.

I wanted the US to do the best science in the world, and that meant we needed to get more talented women into science. Many Americans thought affirmative action would spoil science. But actually, more women = better science. That's because the best candidates taken from 100 percent of the population will be better than the best candidates taken from just 50 percent of the population—and so far, we'd tapped only the best *third* of the United States, the white male part. More women and better science: It's not either/or. You can't have one without the other. The question was how to have both.

We women had been telling each other for years that the only way our needs would be addressed was if we controlled the funding. But experience told us we'd also need an institutional base of support within science and allies (both women and men) to back us. Without those, we'd be back to working around the barriers in our way instead of tearing them down.

When I told the admiral, "No woman, no money," he looked surprised—and concerned. The navy was still recovering from the Tailhook scandal seven years before, when more than one hundred naval officers and Marines—some sporting T-shirts that read "Women Are Property"—were said to have sexually assaulted eighty-three women and seven men at an alcohol-drenched convention in the Las Vegas Hilton. The admiral was wary of more bad press.

"Look, we don't need headlines," I said. "She's chief scientist. That's settled. How do we get her on that submarine?"

The oceanographers would be on an important mission. The US Navy had just completed seven nuclear sub cruises that had provided essential first evidence of the Arctic Ocean's warming. The eighth and final cruise—the one on which Margo Edwards was slated to oversee research—would be crucial, producing "important implications for global warming," the navy predicted.

Why did Edwards need to be aboard the submarine to manage her team's research? "Because when you're out there," Edwards later said, "you can adjust your plan if things aren't going well. Every time I've ever been out to sea . . . you just understand what's going on better. You're there, and you're watching the data coming out."

So Admiral Gaffney and I compromised. The navy would fly Edwards to the Arctic, while her male second-in-command accompanied the team's instruments on the nuclear submarine's trip north. Once in the Arctic, Edwards would get thirteen days to do research under the sea ice in the USS *Hawkbill*.

During the days she spent aboard the *Hawkbill*, Edwards made critically important discoveries about climate change: evidence of thinning ice, volcanoes erupting from the Arctic seafloor, relatively warm Atlantic Ocean water moving into the Arctic Ocean and contributing to the ice melting, and evidence that until about twelve thousand years ago, immense ice shelves up to 1 kilometer thick and hundreds of kilometers long had existed for 2.5 million years. She estimated that the US Navy's six nuclear submarine cruises to the Arctic between 1995 and 1999 increased the world's store of scientific data on the region more than a hundred thousand times. As she said, "You can't study any of that if you're not underwater up there. The ice gets in your way."

Nature published Edwards's report, and Admiral Gaffney graciously wrote the University of Hawaii a letter that, Edwards believes, got her promoted to professor.

I was pleased. I'd helped a woman scientist lead a navy submarine research cruise without incident or even drawing attention to that fact. Some may think my approach was milquetoast, but I've always believed that if you want to get things done, you don't poke sticks in people's eyes. A strategy of using both the carrot and the stick works best—and even better if it's carried out with diplomacy. It's a type of guerilla warfare where you wait to make your move until you can use the system to make things happen for the good of all parties. Then the next time a woman in a similar situation has a problem, you have an ally you can ask for help.

That happened sooner than I expected. Six months later, at an NSF-sponsored research station at the South Pole, physician Jerri Nielsen discovered a lump in her breast. With live video feeds to an oncologist and a cell cytologist in the US, Nielsen performed a biopsy on herself. The international media went into a frenzy for more details on the story, and on July 10—in the middle of the polar winter, when airplane fuel turned to jelly—the NSF air-dropped chemotherapy materials onto the ice. As Nielsen's tumor grew, the NSF organized a hair-raising rescue by the Air National Guard in what was, at the time, the coldest landing ever made on the South Pole. It made for great drama and excellent public relations for the Air National Guard and the Air Force, which insisted they *had to* interview and photograph Nielsen on her way home; it was tradition, they claimed. But for reasons that many women (and men) can appreciate, Nielsen was adamant that she did not want to deal with the media on her way home. The NSF supported her decision, and I made a phone call that would have been difficult if I'd been dealing with the fallout of a public fight with the military over its treatment of women oceanographers. Easily reaching a high-ranking air force officer I knew, I asked quietly, "General, can you imagine what the public will say when the *Washington Post* calls tomorrow about Dr. Nielsen, who will likely speak bluntly about being required to give an interview?" After the briefest of pauses, the general replied smoothly that the air force officers "must have been mistaken about the necessity of an interview." That was the end of that. Nielsen returned quietly to the US for treatment, and when she was ready, she told her story in *Ice Bound: A Doctor's Incredible Battle for Survival at the South Pole*. Susan Sarandon portrayed her in the movie adaptation of the book.

• •

My path to the National Science Foundation, and to being able to help women like Margo Edwards and Jerri Nielsen, began decades earlier when I launched two research organizations in Maryland. The first was the Sea Grant College Program at the University of Maryland in 1977. Congress has funded agricultural research in our landlocked states since the Civil War era; a century later, it allocated a large sum to give the same research opportunities to state universities along the coasts of the Atlantic and Pacific Oceans, the Gulf of Mexico, and the Great Lakes.

Michael J. Pelczar Jr., the leading microbiologist at the University of Maryland, author of the top-ranking microbiology textbook at the time, and vice president for research at the university, worked hard to get a Sea Grant program in our state. Because I had done extensive research in the Chesapeake Bay, Pelczar, who would become a good friend in the years that followed, appointed me to be its first director. The Maryland Sea Grant College studied seafood production and coastal issues such as fisheries, marshes, hurricanes, land loss, rising sea levels, storm surges, coastal populations, and how to mitigate oil spills. Talking with fishermen, oystermen, and area residents gave me practical lessons in the value of scientific outreach to the public, and later, the Reagan administration's attempt to ax the program gave me a useful introduction to the art of lobbying Congress.

While organizing the Maryland Sea Grant College, I made an important friend: Barbara A. Mikulski, a power player in the politics of building connections. Mikulski, the daughter of a Baltimore grocer, began her political career by organizing her working-class neighborhood to defeat a planned eight-lane highway that would have ripped the community apart. From there, she won seats on the Baltimore City Council, in the US House of Representatives, and eventually in the Senate. Mikulski, a Democrat, became the longest-serving woman in congressional history and, as of this writing, was the only woman to chair the Senate Committee on Appropriations. She began her career in the era when, as she tells the story, physician Edgar F. Berman, then widely regarded as an authority on women's health, said women couldn't be leaders because of "raging hormonal imbalances."

When I met Mikulski in the late 1970s, she was chairing the oceanography subcommittee of the Merchant Marine and Fisheries Committee in the House of Representatives. The committee was looking at ways marine bioscience could clean up the environment, and Mikulski had discovered that I was trying to do the same thing at the University of Maryland College Park. From then on, whenever I had questions about a politically complicated situation or needed to move a project forward, Mikulski and her staff always made time to advise me.

"When you're a firstie," Mikulski explained recently, "you want to be the first of many. You want to keep the door open for everybody else to follow. That was our commitment. When we saw talent, we wanted to promote it." Thus, when a Republican president appointed Bernadine Healy head of the NIH, Mikulski and other Democratic congresswomen helped convince the Senate to confirm her.

Another woman in Congress, Constance Morella, became a personal friend and supporter. Morella, like me, is the daughter of Italian immigrants and grew up in Massachusetts. When her sister died of cancer, Connie and her husband adopted her sister's six children and raised them with their own three. Formerly a professor at a local community college, Morella, a moderate Republican, represented my home district in the US House of Representatives from 1987 to 2003. She became chair of the House Science Subcommittee on Technology and later served as US ambassador to the Organisation for Economic Co-operation and Development (OECD). It's no coincidence that Senator Mikulski and Representative Morella were two of the congressional leaders who fought to get the NIH to add an Office of Research on Women's Health in 1990. They both knew what could happen when women are not at the table.

• •

Several years after the Maryland Sea Grant College was up and running, John S. Toll, one of the master builders of America's public universities and a truly honorable man, was recruited as president to the University of Maryland after increasing enrollments almost tenfold at SUNY Stony Brook. Impressed with my work with the Sea Grant College program, President Toll appointed me academic vice president of the University System of Maryland. As a result, while running my lab and traveling to Bangladesh once or twice a year to do research there, I was also cutting my teeth on fractious matters like faculty hires, promotions, and educational programs in Maryland's eleven-campus system.

As long as the men didn't realize what I was doing, I could ensure that strong female candidates were given a fair chance to compete for leadership opportunities at the university. I disapprove of using gender as the sole criterion for personnel decisions, but qualified female candidates can bring unique insights to tasks at hand. Being an éminence grise isn't easy, and knowing when to act isn't magic. They're both hard work. You have to know what levers to pull and when. What we women learned by bitter experience, modern research has confirmed: men who openly help highly qualified women and underrepresented minorities are rewarded, while women and nonwhite minorities who do the same are penalized.

My role as academic vice president also gave me a say in how the university could pursue new areas of research. Revolutions in technology have always fueled explosions in scientific knowledge, often in biology. I'd been watching the Boston area use genomics, computational biology, and bioinformatics to develop world-class centers for biotechnology and genetic engineering. I thought Maryland—home to the NIH, the FDA, and other federal agencies—could do the same. I talked with Toll about how biology was the science of the future and a generator of economic growth. The University of Maryland, I said, should become a leader in bioscience. Toll agreed enthusiastically and became my mentor in institution building. However, after he convinced the governor of Maryland and the state legislature to add a million dollars for biology at the university system's flagship campus in College Park, the campus ended up distributing the money to engineering and the physical sciences. I concluded that the biosciences were not going to be advanced by the usual processes.

The next year, I suggested to Toll that we establish a biotechnology institute. Three key supporters of the project were women: Barbara Mikulski, Constance Morella, and Nancy Kopp, then a leader on the Maryland House budget committee. Kopp later chaired the Maryland House appropriations committee and was elected state treasurer. It was she who taught me how to work with the state legislature and helped me be comfortable asking lawmakers for funding for science and education programs.

During the eleven years I was its president, the University of Maryland Biotechnology Institute received more than a hundred million dollars in state and federal funds for building laboratories and for grants to do research. At its peak, the institute comprised seven hundred scientists and staffers. The experience taught me that I could make a difference—though not necessarily a lasting one, I would come to learn when a later university administration dismantled the institute and distributed its budget and faculty piecemeal to various campuses around the state. Perhaps the most potent lesson I learned from that disappointment was how critical it is to establish internal consensus when building a new institution, so that it's built to last.

Had the university continued to support the institute, the University of Maryland could have become an international leader in biotechnology. Still, the presence of federal agencies, active industrial development, and the outstanding faculty of the University of Maryland campuses today make the state of Maryland the third most important region in the country for biotechnology and the University of Maryland College Park is ranked in the top echelons of research universities worldwide.

••

In 1983, the year I was named academic vice president of the University System of Maryland and cholera researchers were continuing to argue over my discoveries, President Ronald Reagan was having political problems of his own. Reagan had promised to recruit more women to the federal government, but when only 42 of his first 367 administrative appointments were female, women on both sides of the aisle complained. I became a beneficiary of Reagan's attempt to make up for his lamentable record. Nominated by university president Toll, I was appointed to the National Science *Board*—not to the National Science Foundation itself, but to the body that advises the foundation on science policy, funding, and direction. I soon saw firsthand what happens when women lack the power of the purse.

While I was a member and later as chair of the board's polar science committee, Congress doubled its funding for science in the Arctic and Antarctica at the request of the military, as these regions of the world were gaining military and industrial significance. I strongly supported updating science facilities in the polar regions. Antarctica is the least polluted place on Earth, and it's possible to see there what life was like before industrialization. But I was disappointed that there was no similar groundswell from influential women scientists or the voting public to help women of all ethnicities and other people of color in science. If

I'd lobbied publicly for them, I knew many men would have discounted whatever I said on other issues. Besides, the board advised, rather than managed, the NSF and the agency's budget.

In meetings, I soon learned that the board's executive committee made the key decisions. When I expressed an opinion on an issue before the board and a male member said, "You don't have to worry your pretty little head about that," I was the only person who took offense. No one was ever explicitly insulting, but during sessions of the full board, I'd make a suggestion that would be followed by silence or more rarely by a single word—"interesting"—the implication being that my comment hadn't registered. Moments later, a male member of the board would make a recommendation similar to mine, and suddenly the response was "good idea." This was the mid-1980s, and there wasn't a single woman scientist in the country who was a member of the scientific glitterati.

• •

Serving a term on the National Science Board taught me that besides doing science and spending time with my family, I was happiest building institutions. I didn't care much for critiquing others' work, and I wasn't interested in the actual construction of brick-and-mortar structures. I wanted to create institutions to solve problems and make people's lives better for generations to come. I thought of such institutions as "people structures." By now I was working two careers simultaneously: one in scientific research and the other in scientific administration. This is not something I would recommend to everyone. Friends used to call me an "Energizer Bunny" because, with six hours of sleep a night, my batteries never seemed to run down. But I think what fueled me was a profound belief in possibilities, being optimistic that things can be changed and realizing that—with a community of supporters and coalitions—we'd have the power to solve problems. Women scientists had not yet learned to form coalitions to work publicly to reform science. But we were learning that if you studied the facts, identified the problem, picked a long-range goal, and believed in what you were doing, you could accomplish what might otherwise seem impossible. All you need to do to birth a movement is find people you can trust to do a good job, convince them the project is important, give them the responsibility they'll need, and then step back and let them work.

Still, there's no way I could have accomplished all I did without my husband, Jack, and Alison and Stacie, our two wonderful children. I think that if you're surrounded by people who really matter to you and believe in you, you're not only lucky, you can conquer the world. I always had a nurturing place to retreat to, a place where I could recalibrate. Jack would listen to my grumbles and remind me to look at the other person's perspective. When we were not sailing on the Bay we'd go bicycling or hiking or attend a play or concert. One evening when the girls were in grade school, I came home and was trying to get through a pile of housework. I was speaking stridently and being commandeering, when Jack said, "Ricki"—which is what he always called me—"Ricki, this is not the lab. This is home."

• •

My big chance in government came after Congress confirmed Clarence Thomas as a Supreme Court Justice, despite law professor Anita Hill's testimony that he had sexually harassed her. When angry voters sent record numbers of women to Congress in 1992 and Bill Clinton to the White House in the next general election, Clinton appointed a flood of women to high office. In 1997, President Clinton's science advisor asked me if I would be interested in becoming the NSF's deputy director. The White House wasn't thinking about Rita Colwell, scientist, of course. They were thinking of Rita Colwell, science administrator, institution builder, and fund-raiser. In any case, my time on advisory boards had taught me that to accomplish anything, one had to be in charge. I'd already turned down a previous administration's invitation to serve as assistant director of the NSF's education directorate. So this time I said, "No, thank you, but I would be honored and pleased to serve as director if that possibility arose."

One day the following January, my secretary came into my office at the University of Maryland Biotechnology Institute and said, "The vice president is on the phone."

I was busy. "Vice president of what?" I asked.

"Of the United States," she said.

Al Gore was calling to ask if I'd consider becoming the NSF's director. This time I said yes.

I love the National Science Foundation. I believe it's the best agency in the entire federal government, and its staff is unequivocally the best in the world. My six-year term as director, from 1998 to 2004, was one of the best times of my life. The NSF is an unusual government agency because it has no laboratories of its own. Instead, it funds half of the nonmedical scientific research done at US colleges and universities. The NSF is responsible for distributing billions of dollars a year to support the physical, earth, life, social, and behavioral sciences; engineering; and the education of the nation's future scientists. At the time, NSF grants supported thousands of scientists, engineers, teachers, and students each year. I was the NSF's first female director.

The NSF was the outgrowth of an idea President Franklin D. Roosevelt had toward the close of World War II. To win the war, the US had created new defense industries, especially in scientific fields like electronics, but the end of the war meant many of these industries were no longer necessary. FDR worried that the fifteen million uniformed personnel returning to the US after their deployment would find that the jobs they expected to take up had ceased to exist. His wartime science advisor, Vannevar Bush, proposed investing in basic research driven by the curiosity of the nation's scientists, and in 1950, President Harry S. Truman signed the bill creating the NSF, the manifestation of Bush's proposal. Not all discoveries funded by the NSF would have immediate applications, but some of them would create new enterprises and jobs. Today, many economists believe that more than half of the country's economic growth is the result of our government's investment in basic research.

Even before I started as director, NSF old-timers warned me I had several strikes against me. The organization's legal counsel advised me to keep in mind that government's first truism was "IBHWYAG" ("I'll be here when you are gone")—meaning career employees outlast short-term presidential appointees like me and can undo what we have done. Second, I was reminded that I was female. Thankfully, within the agency, no one raised the issue—although early in my term, Jeffrey Mervis, a reporter for *Science*, phoned and asked, "Are you hiring only women?" No, I told him. I wasn't giving preference to women. I was just hiring the very best people for the job.

My third disadvantage was that I was a biologist—the first microbiologist to direct the NSF and only the *second* life scientist to hold the post in the agency's fifty-year history. (A biochemist named William D. McElroy had been director thirty years earlier.) Traditionally, the NSF supported the so-called hard sciences—physics, chemistry, astronomy, and engineering—that had helped the Allies win World War II, while the NIH funded medically related biology research. NSF's directors were typically male physicists or engineers who supported big projects like particle accelerators and telescopes used by astronomers. These male-dominated fields had more prestige than microbial ecology and molecular biology, my fields of study. And yet the biological sciences were exploding with new discoveries and exciting but insufficiently funded research into the ecology of disease transmission, genomics, climate change, global ecosystems, and neural computation.

I had some advantages, however. Senator Mikulski, Representative Morella, and the Maryland congressional delegation were in my corner, and during my time as director, Congress generally liked the NSF because it catalyzed innovation and created new jobs. Information technology, the internet (and, by 2000, all the major search engines used on the web, including Google), the genomic revolution, MRIs, lasers, biotechnology, and nanotechnology are among the foundations of modern life that got their start in NSF-funded university laboratories.

During my term as director, support for science was nonpartisan and apolitical. Neither President Clinton nor President George W. Bush forced any program on the NSF or eliminated any program for political reasons. I liked and worked with people on both sides of the aisle: Ted Kennedy on the left, for example, and Trent Lott, Dick Cheney, and former Speaker of the House Newt Gingrich on the right. I knew Speaker Gingrich socially and enjoyed spending time with him, despite our differences on many issues. When he told me he was fascinated by science, I invited him to give a talk at the NSF about science policy and funding, and he said to call him if I ever needed help getting a budget passed. Later, after Speaker Gingrich had left Congress and President George W. Bush had taken office, there was talk of cutting budgets. The NSF's budget had increased considerably under Bill Clinton, and it was beginning to look as if the new president would give the NSF only a small increase. I phoned Gingrich, he spoke with the White House, and the agency ended up receiving a substantial raise. It wasn't the 13 percent Clinton had recommended, but it was still significant—about 9 percent—and much appreciated.

Another advantage I had was NSF's superb 1,500-member staff. When I first met legal counsel Lawrence Rudolph, he warned me that the two easiest ways to run afoul of Congress or the media were to make a mistake on a personal expense account or to say something in an email that you didn't want to read in the *Washington Post*. I immediately asked Rudolph to check my expense sheets before I submitted them, and I tried to keep my emails impeccably unprovocative. Even my children noticed how circumspect I became in personal emails and phone calls. The wisdom of Rudolph's advice became clear when Representative Jim Sensenbrenner, a powerful Republican who wanted Congress to exercise more control over the NSF, asked for every expense form filed by any NSF staffer and five years' worth of agency travel records. NSF staff spent an entire weekend finding, photocopying, and boxing twenty-five cartons of records, which were then delivered to Sensenbrenner. The NSF heard nothing further from his office.

Decades of experience as a woman in male-dominated organizations had also taught me a thing or two. After sitting through meeting after meeting listening to men ignoring women's voices, I knew we women were expected to remain silent. Supreme Court Justice Ruth Bader Ginsburg recalled that when she was the only woman on the court, she'd make a statement that would go unremarked upon until one of the other (male) justices made the same point—and, as Ginsburg put it, "I don't think I'm a confused speaker." Speaker of the House Nancy Pelosi reports the same. The phenomenon is so entrenched that Denise Faustman, an authority on autoimmune diseases at Massachusetts General Hospital, often took a male student into meetings to make her points. The group found it easier to credit him than her. Sad to say, this problem still exists. A few years ago, during a grant meeting with colleagues I've worked with for years, I made a comment that was ignored. A few minutes later, when a man said the same thing, the point was acknowledged and accepted. Recognizing what had happened, women at the table began emailing "LOL" back and forth, chortling in secret. Of course, none of us called him out on it.

Chairing academic meetings at the University of Maryland, I developed a technique for avoiding interminable discussions that never seem to come to a decision. The technique is, curiously, an outgrowth of my childhood annoyance at not being heard. Unfairness and injustice still anger me, but now I try to put that experience to use by making sure everyone has a chance to say what they're thinking, especially those who are uncomfortable speaking out unless they're asked to do so. That system ensures that I'm being fair, too.

First, as a meeting gets under way, I make sure that even quiet committee members contribute their ideas—preventing any one person from taking over the meeting and squashing dissent. (Interestingly, a team led by Dr. Anita W. Woolley at Carnegie Mellon published research showing that the effectiveness of a group depends on the proportion of females in it, because women will listen to everyone in the group, not just to the few people hogging the conversation.) Then I identify gaps in our discussion of the topic at hand, articulate a workable conclusion, wait for someone else (inevitably male) to restate what I've just said, listen to his colleagues say, "Brilliant idea!"—and finally ask for a motion for committee action on this brilliant idea. You don't get the credit in that moment, but you will when the job gets done. If you're concerned about immediate praise, you risk achieving nothing.

In addition to managerial strategies honed over twenty years in science administration, I arrived at the NSF with two other important advantages. First, of course, was Jack, who'd retired in 1989 after a career as a physicist at the National Institute of Standards and Technology to run the household—and sail. Jack had decided that he really wanted time to do things he loved, not just sailing, but also biking, amateur astronomy, and monitoring his favorite pair of eagles nesting in a tall tree across the Potomac River from our home. Alison and Stacie were in graduate school and medical school, respectively, and we were financially comfortable, thanks to Jack's careful planning. So his retirement was logical and welcome. My second advantage was my former graduate student Anwar Huq, who, with his wife and children, have become members of the extended Colwell family, so much so that when Anwar and Jack were hospitalized at various points in their lives, each sat at the other's bedside. Anwar is kind, honest, smart, and extremely knowledgeable about the practical aspects of microbiology. He kept our lab running all the time I was at the NSF.

• •

And so, with all these people and skills in my back pocket, I arrived at the NSF with a list of goals and a clear strategy for achieving them. My immediate goal was to help graduate students and K–12 science education. Stipends for graduate students in science and engineering were so low that many students were dropping out or going on welfare and food stamps. I wanted to boost NSF's $14,000 graduate research stipends to $30,000, and did so by starting a new fellowship program that paid graduate students to bring their state-of-the-art research into public school science classrooms five hours a week. Grad students would learn how to teach, and kids would benefit from having young men and women scientists and engineers as their instructors. We called the program GK-12. An important by-product was that the traditional NSF graduate research stipends were also increased to $30,000. Unfortunately, a later NSF director axed the GK-12 program, and graduate research fellowship stipends have stagnated again.

In the longer term, I knew the director's job was "budget, budget, budget." By the late 1990s, the NSF could fund only 9,000 of the 32,000 grant applications sent to us by scientists each year. I wanted to double the NSF's budget and, thanks to my previous experience serving on various boards and committees, figured I knew how to do it: consensus. When everyone is working with you—right alongside you, not far out in front of you pulling things apart, or so far behind that you have to haul them along with a rope—things can get done.

The NSF had to set priorities. If we didn't, Congress would set them for us. I also knew from working with the Maryland legislature that if I went to Capitol Hill and asked them to double our budget across the board, we'd get only small, incremental increases—a few percent a year at best. Besides, I'd never been fond of what I call the peanut butter method—spreading funds evenly and thinly; I believed good scientists should get the money they needed to do a good job. I was also adamant that you shouldn't take funds from Peter to pay Paul. Never start a new program by reducing the funding for important research that's already under way. New initiatives should be used to attract substantial new funding.

But what new areas needed funding? We already knew several fields of science and engineering were receiving far more proposals for good research projects than we had funds for. Each of those areas represented a lost opportunity for the US economy to grow. And so, over the six years of my term as director, we worked on establishing a series of new programs. Thanks to the years I'd spent switching from one specialty to another in my career, I knew we were moving into a new era where some of the richest problems in science lie where traditional specialties overlap. As I explained to a Senate appropriations subcommittee, "Up to now, we have sought understanding by taking things apart into their components. Now, at last, we can begin to map out the interplay between the parts of complex systems."

In 1998, convincing physicists, biologists, astronomers, computer scientists, mathematicians, engineers, and social scientists to design, carry out, and analyze experiments together—from beginning to end—was difficult. Dissolving the artificial boundaries between academic departments also proved controversial. Disciplines, ensconced in separate departments like little fiefdoms, preferred to control their own funds; sharing was viewed as giving away precious resources. There remains a sense that interdisciplinary research is somehow less rigorous, and that anything that's never been done before can't be good.

The NSF itself had long operated within silos, with proponents of the physical sciences separate from those overseeing the life sciences, who, in turn, were separate from those advocating for the social sciences. Many scientists actually wanted to join interdisciplinary teams—requests for grants for nanotechnology, in particular, were pouring in—but when I became director, the NSF wasn't structured to fund such research.

Nanotechnology involves biologists, engineers, physicists, and chemists working together to manipulate matter, one atom at a time. When my predecessor, Neal Lane, left the NSF to become science advisor to the White House, he and I worked together to obtain funding for this new and exciting offshoot of bioengineering to solve medical and industrial problems with new methods of manufacturing and new ways of producing pharmaceuticals and delivering treatments for chronic diseases. I was excited by the prospect of biology becoming an economic engine. But one of my proudest achievements at the NSF came from supporting one of the biggest transformations of the 1990s—the computer revolution.

Interdisciplinary teams tend to generate enormous amounts of data, and I knew computers were going to be a game changer for every discipline, from mathematics and education to the social sciences. Yet our country's universities and colleges had unequal access to high-end computers. Unless we could computerize and connect all our scientists, they wouldn't be able to share or compare their data. So, after a long discussion, the NSF agreed to ask for a *billion*-dollar program to bring high-end computing to all our academic institutions. To do that, we had to bring funding for the new field of computer science up to par with the funding given to physics and engineering—without alienating the old-timers. I invited the heads of all the relevant government agencies to a meeting in my office to get their support. They agreed the NSF should take the lead, although we'd all work together.

I recruited a University of Pennsylvania robotics professor, Ruzena Bajcsy, to spend three years at NSF as assistant director of our computer and information science and engineering (CISE) directorate. Bajcsy was born in Czechoslovakia in 1933, the year Hitler became chancellor in Germany. Her parents were Jewish and, although they had converted to Catholicism, an anti-Semitic maid had murdered her mother when Bajcsy was three years old, and Nazis killed her father and stepmother when Bajcsy was eleven. After what she'd faced, mathematics and computers were "comforting," Bajcsy told an interviewer. "People are unpredictable. With machines, you know what you get," she said, "and if you don't get what you want, you know it's your fault! So somehow you have control."

After earning degrees in electrical engineering from Slovak Technical University and leaving her family behind, Bajcsy arrived in California on a one-year fellowship to Stanford in the middle of 1967's

"Summer of Love." "I came from this very conservative Czechoslovakian, communist, and in some ways very Catholic background," Bajcsy recalled. She discovered what freedom was all about the day she visited the Stanford bookstore and standing next to each other on the same bookshelf were three books: the Bible, the Koran, and *The Communist Manifesto*. When the Russians moved into Czechoslovakia the following year, Bajcsy opted to stay in the US, not realizing that it would be fifteen years before she would see her two children again.

For the first time, Bajcsy found herself in an environment that didn't fear new ideas. "The whole area of computer science was still in diapers, so to speak," she said, "and artificial intelligence was just being born." Bajcsy would spend thirty years at the University of Pennsylvania's engineering school. By the time I met her, she had two PhDs, one in electrical engineering from Slovak Technical University and another in computer vision and pattern recognition from Stanford. Bajcsy is now a member of both the National Academy of Engineering and the National Academy of Medicine. She's a star.

At the NSF, Bajcsy soon discovered that being a program manager involved competing for money to fund grants for your particular field of science. Physics and engineering had always gotten the bulk of the NSF's budget, so even in years when Congress gave the agency an incremental budget increase of 1.5 percent, they knew they'd be fine; smaller fields like computer science would not. Gathering all the program managers together, Bajcsy disarmed them by passing around cheese and wine like their grandmother as she gradually sold them on our vision for computerizing all of science.

If we gave all our new money to computer scientists, I knew we'd exacerbate the jealousies of academia. So I insisted on a bit of guile: we'd give the bulk of our new funding to computer science and engineering, but enough to all the other sciences to accustom them to using computers. Everybody got a piece of the action to incorporate computer science into every discipline. And it worked. No one stood on the sidelines enviously eyeing computer science with its newly granted millions of dollars. Instead, every discipline became part of the computer revolution.

Mathematics was another interdisciplinary priority. Mathematics is the lingua franca of every science, as well as engineering, business, and the social and behavioral sciences. Some of the greatest insights into contagious diseases like AIDS have come from mathematical models, and math is the lifeblood of the IT revolution. But the United States' leadership in mathematics was in decline. The NSF is the major source of grants for educating PhD students, but math funding had increased only 1.5 percent a year, less than the rate of inflation, for a decade.

In 1997, a former head of the National Security Agency, Lieutenant General William E. Odom, circulated an influential report about our desperate need for more homegrown PhD mathematicians. He and many members of Congress disapproved of our dependence on immigrant mathematicians and experts from the former communist bloc. I was already committed to requesting double the NSF's math budget over four years, the payoff being 45 percent more students earning math PhDs.

Another interdisciplinary goal was especially close to my heart. I've always seen the world as a nicely woven quilt, all of a piece. I wanted the NSF to bring all our scientific tools and methods—from mathematics, physics, chemistry, ecology, biology, and social science—to explain how the natural environment and its animals and plants interact in complex holistic systems. Trying to think of a name for the program, I assumed anything involving the words "ecology," "holistic," "diversity," or "environment" would be dead on arrival in Congress. "Oh, it's so complex," a biologist complained. "Okay," I said, "let's call it biocomplexity."

Names can be important—even a seemingly incomprehensible name like "biocomplexity." Without such names, nothing gets done. Legislators need names they can use to explain to voters how a program will produce new industries with new jobs. We had already changed one name for appearance's sake: STEMM was originally called SMET, but to a bacteriologist like me, that sounded too much like *Mycobacterium smegmatis*, a bacterial species commonly referred to as *Mycobacterium* "smet" that is associated with human genitalia. So we rearranged the acronym.

Eventually Congress asked me what biocomplexity was. I explained with the example that if they wanted to build a new highway with maximum human benefits but minimize the financial cost, biocomplexity would give them the tools to combine information about the region's watershed and water table, census data about where the population was growing or shrinking, and relevant details about the area's geology, flora, and fauna. "That's biocomplexity," I said. We got the money. And the spirit of biocomplexity lives on in programs run jointly by several NSF departments—a new management practice for the agency.

• •

Consensus was the name of the game. That is, until we tried to help women of all ethnicities and increase the numbers of other underrepresented minorities in science.

A decade before I came to the NSF, a woman there issued—on her own—what she called "the memo that changed the world." Mary E. Clutter was in charge of the NSF's biological, behavioral, and social science funding and had served twice as the agency's acting deputy director. The experience had taught her that, as she said, "if you want to do something, just do it." So one day in 1989, she announced that the NSF would no longer support conferences or workshops in the three subjects she was overseeing if they had no women speakers and no explanation for the omission. "It's a matter of including women every damned time," she told *The Scientist* magazine. Clutter knew the memo would be unpopular. "Men would call in and say, 'I'll report on my wife's work. Isn't that okay?'" she later recalled. The NSF's then executive director, W. Franklin Harris, backed her up. Women were already getting 33 percent of PhD degrees in the life sciences, and "for those fields," he wrote in *The Scientist*, "it would be only under the most extenuating circumstances that conferences, meetings, or international congresses would not be able to identify qualified women as invited speakers." Sadly, Clutter's memo did not change the world. She had no power to enforce her words. And even if she had, she would have run afoul of congressional overseers eager to pounce on anything with even a whiff of affirmative action.

Ten years later, when I was director and Clutter was my assistant director for the biosciences, the nation was bitterly divided over helping women and girls in science. Programs to support women in science and engineering that the NSF had quietly run for decades were getting caught in a legal quagmire. In this fraught atmosphere where so much was at stake, consensus seemed impossible.

The 1990s were different from anything we had ever experienced. Women scientists in my generation were isolated pioneers who'd managed, against all odds, to find a way around—or through—closed doors. The generation that followed benefited when pressure from the feminist movement pried open some doors in the 1970s and '80s. But during the 1990s, large numbers of women were throwing open the doors of science, engineering, and medicine, only to find there were too few senior women faculty members to be their mentors or whose research teams they could join. Women weren't being promoted to full professorships and, once again, were dropping out or getting stuck in low-level jobs. Since it costs $1 million to train a PhD scientist, society loses an astounding amount of money whenever a woman with a doctorate drops out of science. Women needed the same kind of career opportunities that white men had always had. Women needed an institutional power base within science.

Although few voters demanded action on the issue, many Republicans and Democrats in Congress and the White House did. The US wasn't educating enough skilled STEMM workers to meet the needs of our economy, and our dependence on scientists and engineers from other countries had become a national security concern. Women constituted our biggest untapped source of talent. In 2000, Congress passed a Science and Engineering Equal Opportunities Act, calling for a comprehensive government program to increase the number of women and minorities in STEMM. So far, so good. But after that, the consensus fell apart.

As affirmative action opponents mounted legal challenges in federal and state courts and legislatures to a variety of programs, the NSF found itself walking a legal tightrope. In 1997, Travis Kidd, a white mathematics graduate student from South Carolina, sued the NSF for providing certain graduate research fellowships specifically for underrepresented minority students. My deputy, Joe Bordogna, a former engineering dean at the University of Pennsylvania, wanted to fight the case in court. Like me, Bordogna was an Italian American who'd lost a parent early in life and whose mother had supported the family with a miserable factory job (gluing labels on whiskey bottles, much as my mother had glued the soles onto shoes). One of the reasons I'd asked him to be my number two was that we both believed, heart and soul, that university cultures had to change; if they did not, women of all ethnicities and African Americans, Latinos, and other underrepresented minorities would never be recruited, retained, or promoted. Yet US Justice Department lawyers told us we'd lose against Travis Kidd's suit, jeopardizing all our other programs for women and minorities, including STEMM camps for girls and grants to individual women scientists. Reluctantly, the NSF agreed to settle the case and pay Kidd $14,400 and his lawyer $81,000. Nervous university legal advisors began cautioning universities not to mention the words "race" or "gender," and three of the four governmental agencies responsible for ensuring equal rights for women in federally funded education stopped defending Title IX.

I was reminded of the legal obstacles women had faced when I was just starting out forty years before: the state laws and university regulations against giving women equal access to jobs and training. With the way forward blocked once again, we would have to do what we'd done in the past: find a way around, over, or under the barrier. I didn't know what else to do.

Convening a special panel of NSF program directors, I told them I wanted to make the status of women in academic science a national priority. This would require transforming the misogynistic culture of universities. And I needed them to find a legal way to do so. "Show me how to do this," I said.

So they did.

The solution was deceptively simple. We'd create a new type of grant: both men and women—"one hundred percent of the human race," as a university lawyer joked—could compete for the grant, provided their goal was to raise the status of female assistant, associate, or full professors in science and change the culture of their universities. (I'd wanted to include postdoctoral fellows in the program, too, but not all fields have postdocs, so that part of the plan was scrapped.) The grants were large, ranging from $2 million to $5 million each, spread over five years.

But how would such a grant work? Bordogna and I argued about many things during the six years we worked together; as he put it, "We're both Italians!" But we came to serious loggerheads only once. As a former dean, Joe believed any university dean, provost, or president would leap at the chance to reform their institution. My experience with university administrators told me they wouldn't. I thought giving big grants to substantial numbers of outstanding female scientists would show biased academic administrators that women could do excellent science. Rather than go down fighting, Bordogna and I compromised. We'd split the grants: half to individual women and half to top-level administrators. We called the program ADVANCE and folded it into a package of programs with a Congress-friendly name: 21st Century Workforce. Alice Hogan, the program director, came up with the idea of having each university president, provost, or dean (almost all of them male) personally sign on the dotted line for the money, thus accepting responsibility for its results. I was skeptical at first, but within a few years, my idea of relying on big grants for individual women scientists was quietly dropped because Joe's plan was clearly more successful.

Between 2001 and 2018, the NSF spent $270 million contracting with more than one hundred institutions of higher education to design programs to solve specific problems experienced by women in science, technology, engineering, and mathematics on their campuses. Not every university did great things with their ADVANCE grants, but by 2011, the first nineteen universities were paying women more equitable salaries, promoting more women to senior management, hiring more female junior faculty, and promoting more women faculty to full professorships. The most astounding outcome of the program was that, according to the NSF's Division of Science Resource Statistics, the number of female full professors in the US increased by 8,939 during that time. Had we taken my original approach, we would have raised the number of female full professors by only 540. Bordogna's way was faster and more productive and remains so to this day. I'm very glad he persisted.

••

For really big change to be possible, we were going to need outside help. One day early in my term as NSF director, when Newt Gingrich and I were in a meeting, I told him, "Speaker Gingrich, the NSF budget must be doubled. It has to be doubled."

"Director Colwell, you're wrong."

My immediate thought was, *Oh dear. This is not good.*

But then he said, "It should be tripled." In another committee meeting, Bordogna heard Senator John McCain say the same thing. And Senator John Glenn, the former astronaut, would testify to a congressional committee that the NSF's budget should be "five times" bigger.

Senator Mikulski and Senator Kit Bond, a Republican from Missouri, were ready to lead a bipartisan effort to increase the agency's budget substantially. Pulling out our list of the critical new areas of science we wanted to support—including mathematics, computer science, ecology, and women's programs—we began informing members of Congress about each one. (I use the word "informing" because it is illegal for agencies to "lobby" Congress. Instead, we "inform" its members about what we'd like to do.)

I'd been walking both sides of the aisle ever since becoming director of the NSF, and assumed I'd be a key player dealing with Congress myself. But the NSF's legislative affairs unit protested. Traditionally, directors didn't go to Capitol Hill to speak to Congress about money. When I discovered the unit was making budget agreements with congressional aides without telling me, however, I insisted on speaking to Congress myself.

Walking the halls of Congress felt like old times in the Maryland statehouse. Congress was a bit more refined—its members didn't munch on their lunches during hearings—but some of the old low-level bias remained. When I testified with Bordogna at my side, some congressmen and their male staffers would direct their questions to my deputy instead of to me. Still, I found that most of the politicians I encountered—from both sides of the aisle—were fundamentally well-meaning. The lesson here is that scientists need to spend time making sure their congressional representatives visit the universities and other science facilities in their districts so they can better appreciate the research that's going on. And of course, it's important to be kind and to show respect. Personal relationships—that's what it's all about.

In the end, I couldn't double the NSF's budget. I did manage to get it increased by 63 percent—extra funds we used for productive and socially relevant programs, including major funding for computing, student stipends, mathematics, women professors, large-scale physics and astronomy projects (one of which confirmed Einstein's general theory of relativity), a new airplane for high-altitude research, and earthquake research centers. I am certain that having Neal Lane, my predecessor NSF director, in the White House Office of Science and Technology Policy during the first two years of my tenure was an important factor in my success with the budget. I had hoped to do more, but as of this writing, this was nevertheless the greatest period of growth in the NSF's fifty-year history. (Details explaining how the budget increase is computed appear in the endnotes of this book.)

When I speak with women about careers in science, I always mention opportunities for women scientists in government, because until women head government agencies and serve in Congress and state legislatures as regularly as men do, real and sustainable change will not occur. Science and engineering will continue to suffer—and the American public will, too. I could see this clearly and firsthand when I chaired a committee looking into the nation's defenses against bioterrorism. The experience would change my life.

chapter seven

The Anthrax Letters

I'm attending a meeting of the CIA's Intelligence Science Board, established after 9/11 to give the director of national intelligence independent scientific advice. Our meetings are secret, so I cannot say precisely what is being discussed. But I can reveal that, no matter what I say, I cannot get the rest of the committee to focus on the risks posed by biological agents like bacteria and viruses, some of which can be more lethal than bombs and explosions—because biological agents, once unleashed, can grow and spread.

The members of the committee are almost all men, most of them engineers, physicists, and chemists from Stanford and the Ivys. Initially, I'm the only biologist on the board and one of only a few women. I'm there not just as the former NSF director but as an expert on bioterrorism. The men aren't rude. They're pleasant and very dedicated. They just don't pay attention to me because I'm not part of the old guard. Even when two other biologists (both women) join the board and a special session is devoted to biothreats, we cannot convince our colleagues to consider bioterrorism a very serious threat.

• •

There have been many examples in this book like the one above, where women aren't heard, or given a seat at the table, or allowed to be leaders. This chapter will tell the story of an extraordinary exception to this rule.

In the tale I'm about to tell you, I was in charge. Maybe it takes an emergency for something like this to happen; it did for me. But without knowing it, for my entire career, I had prepared for the anthrax-laced letters that terrorized this country in the autumn of 2001. Everything in my career had been building up to this point. Never once in our seven-year-long investigation tracing the source of that anthrax did I fear that I didn't know what to do or how to do it. I remember this calm, inner voice saying, *Think carefully and act methodically*. We were men and women from different backgrounds, listening to each other, working together and—thanks to our diverse points of view—solving an immensely complex problem threatening the safety of our nation.

To tell this story properly, we need to go back to the beginning.

• •

My connections to the US Intelligence Community (a group of seventeen governmental intelligence organizations, including, most notably, the CIA) began in the early 1970s, when I was an associate professor at Georgetown University. I was elected president of several international biological organizations that held meetings in South Africa under apartheid, in Czechoslovakia before the fall of the Berlin Wall, and in other politically interesting places. When I wrote articles for scientific journals about the science being done there, the Intelligence Community noticed. My work on cholera also caught their attention. The CIA was interested in environmental problems like droughts and epidemics, which have, historically, created massive political instability. And during the 1980s and '90s, cholera became a problem throughout South and Central America, areas that had not seen a significant epidemic in a century.

My work with cholera got me thinking about something else that interested the CIA: the idea that someone, somewhere, might deliberately use microbes to harm people. I became convinced that the federal government should be studying ways to deal with *biological* weapons, *biological* accidents, and *biological* attacks. I was concerned about the possibility of someone using legitimate scientific research for military purposes or societal disruption.

In the late 1990s, I approached John R. Phillips, a PhD chemist and the chief technical officer of the CIA (who would soon become chief scientist for the entire Intelligence Community) and asked, "Is there a committee addressing bioterrorism?" And when he replied that there were none currently active, I asked, "Shouldn't there be one?"

Months went by, but none of us advocating for a database of pathogens made much headway. Or so we thought. Then, in January 2001, word came that Phillips, both a scientist and a very effective administrator, had secured funding from Congress to begin a major research and development program for countering chemical and biological terrorism. I was pleased to be invited to join the project's advisory board, even though the first meeting of our group could not take place for another nine months.

In the meantime, something terrible would happen.

• •

At eight forty-five a.m. on the morning of September 11, I was in my office at the National Science Foundation in Arlington, Virginia, when a staff member rushed in. A plane had crashed into one of the World Trade Center's twin towers in New York City. Turning on a TV, we watched a second plane slam into the other tower and realized this was no accident. An hour later, we learned that a third hijacked plane had crashed into the Pentagon, a ten-minute drive from our offices. A fourth plane was over Pennsylvania, racing toward the White House. Like many Americans, I felt helpless. It was surreal.

Three weeks later, on Friday, October 5, the committee on bioterrorism held its first meeting at CIA headquarters. With the nation on high alert for post-9/11 attacks, we had a tense but informal two-day workshop about chemical and biological terrorism and possible defenses against it. Although what was said there is still classified, the session was highly productive. But what really sticks in my memory is something I heard afterward.

A press release brought disturbing, electrifying news: a Florida man, Robert Stevens, had died from inhalational anthrax, a dangerous infection of the lungs that quickly spreads throughout the body, often resulting in death. His was the first US anthrax case since 1975 and only the eighteenth case in the twentieth century. Everyone on the committee knew a lot about *Bacillus anthracis*, usually called simply anthrax. They knew that Stevens's anthrax could have come from a rogue nation state, a homegrown terrorist group, or a crazed and isolated individual. Given the timing—so soon after 9/11—I and many others at the meeting and in government assumed al-Qaeda was following up on the World Trade Center and Pentagon attacks with biological warfare.

There was a certain logic to our thinking. Several of the terrorists involved in 9/11 had lived in or learned to fly airplanes in Florida, so it seemed plausible that they might have chosen a Florida resident as their first bioterrorism victim. Besides, anthrax had been used as a weapon during World War I and is still considered one of the most likely biological agents to be adopted by terrorists, as its spores are easily found in nature, can be mass-produced in a laboratory, and can live for decades or longer in the environment. The government had been so concerned about a possible bioterrorism component to the September 11 attacks that it had taken hourly samples of the air around the Pentagon for a week and tested it for *B. anthracis*.

When the president was told about Stevens's illness, he assumed it was part of a biological weapons attack. Publicly, however, government health officials moved quickly to assure the nation that there was no need for concern. Secretary of Health and Human Services Tommy Thompson repeated six times on national television that Stevens's illness was apparently "an isolated case." Thompson suggested Stevens might have become sick from drinking stream water during a hike in North Carolina. The story made the government look uninformed—even clueless. A bit more information about *B. anthracis*'s life cycle will explain why.

• •

Anthrax infects primarily large grazing animals: cattle, sheep, goats, pigs, elephants, and, in Northern Canada, bison. *B. anthracis* spends most of its existence in the soil in a dormant state (think hibernation) within a tough capsule called a spore coat, with which it protects itself from the elements. When an animal grazes on grass growing in contaminated dirt, it can breathe those spores into its lungs. Inside the animal's nutrient-rich body, the spore coat breaks open, releasing the bacterium. The cells multiply explosively, enter the bloodstream, and overwhelm the animal's immune system. Even a healthy animal can die of anthrax in two or three days. When putrefaction or scavengers open the carcass, anthrax cells spray into the air. Deprived of nutrients, they quickly starve and go dormant again. In this cycle of dormancy and infection, they can survive for decades—perhaps even centuries.

Humans generally encounter *B. anthracis* through direct contact with infected animals or products from those animals. A person can become infected with cutaneous anthrax, a relatively mild form of the disease, if they have a break in their skin and touch the unclean hide or wool of an infected animal. Eating infected meat can produce a serious and sometimes fatal illness. But the most feared form of anthrax is inhalational anthrax, the type that killed Robert Stevens. Until recently, inhalational anthrax was thought to kill 90 percent of its victims. Strong public health systems, vaccinating domestic cattle and sheep, and burning all contaminated carcasses immediately to ash have eliminated anthrax among humans in the Western world. We now know that prompt and aggressive treatment with modern antibiotics can save more than half of anthrax victims.

As the first microbiologist to direct the National Science Foundation, I knew I had a unique opportunity to bring my scientific expertise to this problem. All my experience as a microbiologist told me that Stevens's death wasn't an accident—and that unless we could identify the precise genetic makeup of the bacterium that killed him, we would never identify his killer (or killers). Furthermore, the sequencing had to be done soon, before laboratory manipulations unintentionally altered the bacterium's DNA.

Since the NSF funds scientific research, the agency's staffers knew the nation's leading researchers in all the relevant sciences, and the agency was in a unique position to harness their intellectual capital. This would be new territory for the NSF. Its job wasn't to investigate crime; that was for the FBI. But we could perhaps use genetic science to identify the precise biological weapon the murderer had used.

I quickly offered John Phillips and the Intelligence Community the expertise of the NSF—and he immediately accepted, without even a handshake between us. We'd worked effectively together before 9/11. Phillips, Linda Zall of the CIA, and I had been helping to declassify thousands of Cold War spy satellite photographs of Arctic sea ice, which offered stunning early evidence of climate change; they showed the Arctic ice shrinking by twice the area of Texas between 1979 and 2013.

Phillips and I had also created a program to use the CIA's deep pockets to help finance—publicly, and with no strings attached—scientifically worthy projects that the NSF could not afford on its own. Together, we provided the first grants for Google founders Larry Page and Sergey Brin in 1998, when they were still graduate students working on a search engine to link and rank web pages on the internet.

After 9/11, believing that the CIA needed access to excellent science to deal with the aftermath, I invited Phillips to meet with twenty to twenty-five NSF program managers in my office. Phillips and his scientific staff (half of whom were women, by the way) had been living on sugar highs, working day and night on sacks full of cookies his wife baked each day. Information was pouring into the country, and the scientists were assessing its importance for the president and the National Security Council. The NSF could help Phillips connect with leading researchers.

"What are we going to do to help you?" I asked Phillips when he arrived. "What are the problem sets you're dealing with?" Over the course of the next two hours, Phillips heard what each NSF program manager dealt with—and when one of them said, "I'm doing KDD, knowledge discovery and dissemination," he was particularly interested. KDD helps analysts quickly discover information in disparate sources of data.

In the meantime, Phillips had forwarded my offer to help to the FBI. It is hard to believe that, in 2001, the FBI employed only two microbiologists and had no facilities for identifying the entire genetic content of a bacterium or a virus that could be used for bioterrorism. Sequencing the entire genetic material in bacteria seemed far too unreliable for crime solving.

The NSF, however, knew the best scientists for determining the order of every nucleotide pair on a bacterium's strand of DNA. The agency had funded, or was funding, many of them. So we agreed to an unusual working relationship with the FBI. We would share whatever science we learned with the agency, while it would hold details of its criminal investigation as close as possible for use in an eventual courtroom trial. Over the next seven years, the collaboration between the government's scientific agencies, the CIA, the FBI, and the Department of Justice was excellent. It reached a point where a red phone with a direct line to the CIA was installed in my office for emergencies, like something out of a James Bond movie.

While the NSF was talking with the CIA and the FBI, the White House and Congress were getting an education about the serious threat posed by biological warfare. Former White House press secretary Ari Fleischer remembers sitting with George W. Bush through "the most gruesome briefing in the Oval Office about anthrax, how it could spread, and how we had no defenses" against a mass assault. A cabinet officer confirmed what many of us had been arguing for years: the nation was "grossly unprepared for a biological attack." Both the president and first lady were advised to take ciprofloxacin (brand name Cipro), one of three antibiotics approved by the FDA for treating inhalational anthrax. (We learned at the time that if just three American cities were sprayed with *B. anthracis* simultaneously, there would not have been enough Cipro to treat their civilian populations.)

We had no national strategy system for organizing the scientific research and development necessary in the event of a bioattack. The military's science advisors were almost exclusively trained as engineers or physicists and were focused on threats posed by nuclear bombs and other explosives. Accustomed to thinking about radioactivity, they had no experience with epidemics that could kill thousands, if not hundreds of thousands, within weeks. They didn't fully understand that the destructive capability of a nuclear bomb and a weapon armed with *B. anthracis* spores was roughly the same. And that a lone terrorist could make an anthrax weapon much more easily than a bomb.

In the months and years to come, I would use my seat on the CIA's Intelligence Science Board to try to prepare the country for a full range of possible attacks—because if you look only for conventional threats, you'll only find conventional threats. But that October, my priority was tracking down whoever had caused the biological attack that was already under way.

••

Anthrax offered an unusual and extremely difficult scientific problem: *B. anthracis* found in one part of the world shows very few genetic differences from *B. anthracis* found elsewhere. In most bacteria, mutations will occur over time as the organism adapts to new environments. But *B. anthracis* is believed to spend much of its life in a state of suspended animation, during which it does not replicate its DNA. Further, when anthrax does kill, it kills fast. The process from dormancy to ferocious toxicity and back to dormancy moves rapidly. Variations would be rare—and exquisitely hard to find.

I believed the only scientists who could find those differences were J. Craig Venter, Claire M. Fraser, and their team. Earlier that year, Venter and Francis Collins of the NIH had reported the sequencing of the human genome, and *Science* had recently called Claire Fraser "the undisputed world leader in microbial genomics." Both Fraser and I believed we had to do full sequencing—determining the order of every nucleotide pair in the bacterium's DNA—in order get the full molecular fingerprint to distinguish one pathogen from another.

In the fall of 2001, sequencing was still formidably expensive and time-consuming. Most research investigators were sequencing as little as 1 percent of a bacterium's DNA and assuming that that tiny snippet represented the whole. We wouldn't be starting from scratch, though. The Institute for Genomic Research (TIGR), a nonprofit research organization, had already sequenced the full genomes of two bacteria, *Mycoplasma genitalium* and *Haemophilus influenzae*, and was already considering ways to sequence the five million-plus bases comprising the genetic code of *B. anthracis*.

Hurrying back to my office the afternoon I learned about Stevens's death, I prepared to telephone Fraser, who was running TIGR. My position was delicate, though. Legally, NSF's director couldn't offer a researcher money; the agency's experts must thoroughly review each grant application and decide whether to approve it. Waiting until six p.m. so our conversation would be private, I called Fraser. Carefully following the advice of NSF's legal counsel, I advised her that if an unsolicited proposal to sequence strains of *B. anthracis* arrived at NSF, it would get a fair review—*and* we could make emergency funds available quickly, within one or two weeks instead of many months.

Unfortunately, I was so careful that she didn't understand what I was trying to say. Like many Americans who had heard Secretary

Thompson on the news, Fraser thought Stevens's death was "a single crazy incident." So I spent the next week waiting for a grant application that did not arrive.

• •

During those agonizing days, the country learned more about Stevens and his illness. Stevens had been a photo editor for two supermarket tabloids, the *National Enquirer* and the *Sun*. Admitted to the hospital on the morning of October 2, he was almost unconscious and unable to talk. Medics thought he had bacterial meningitis and gave him a spinal tap to confirm the diagnosis. Laboratory technicians called in Dr. Larry M. Bush, the hospital's infectious diseases specialist, for an emergency consultation. Examining the spinal fluid under a microscope, Dr. Bush saw chains of large bacteria that looked to him like anthrax.

Dr. Bush had never seen a case of anthrax. But he had read recent articles in the medical journal *JAMA* about various biological weapons of mass destruction. Concerns about *B. anthracis* had circulated recently after reports that Iraq had a biological weapons program and that a biological weapons factory in Sverdlovsk, Russia, had accidentally allowed anthrax spores to escape, resulting in the deaths of sixty-six people 2.5 miles downwind.

Accompanying *JAMA*'s anthrax article was a color photograph of the bacteria, which, to Dr. Bush, looked to be the same kind in Stevens's spinal fluid. Following procedures established as a result of the Sverdlovsk incident, the CDC notified the FBI. Florida TV news soon showed FBI agents in hazmat suits collecting evidence at Stevens's workplace. A whitish powder containing anthrax spores was found on his computer keyboard and desk and in the building's mailroom. Two mailroom employees tested positive for anthrax; both were treated successfully. Presumably, Stevens and his coworkers had come in contact with the spores from an envelope or package sent by mail, but no such envelope or package was ever found; the business routinely incinerated its trash.

Suddenly, on October 12, a week after Stevens died, the situation changed dramatically. An office worker in New York City opened an envelope addressed to NBC news anchor Tom Brokaw. Inside, she saw whitish-gray powder and a photocopy of an unsigned, misspelled message written in uppercase block letters:

THIS IS NEXT
TAKE PENACILIN NOW
DEATH TO AMERICA
DEATH TO ISRAEL
ALLAH IS GREAT

Three days later, Grant Leslie, a young woman interning in the Washington, DC, office of Senate Majority Leader Tom Daschle, opened a similar letter. Realizing immediately that it might be *B. anthracis*, Leslie called for help and carefully held the envelope at arm's length. Her quick action saved enough powder for extensive testing. More than 625 people who might have been exposed in the Hart Senate Office Building were lined up for nasal specimen swabs and in some cases given antibiotics, swift action that may have saved lives.

More anthrax-laden letters arrived in the coming days, one addressed to the *New York Post* and one to Senator Patrick Leahy. Senator Leahy's letter was recovered unopened, with almost a gram of spores inside, enough to analyze fully. All the letters had been delivered by the US Postal Service. These letters changed everything. We now had no doubt that Stevens had been killed deliberately. We were dealing with terrorism.

As soon as Fraser realized the extent of the crisis, TIGR applied for a grant. The NSF approved the application within a week, unusual speed for the federal government. On October 26, three weeks after Stevens's illness was discovered, Claire Fraser and her principal co-investigator, Timothy D. Read, had funding "to sequence the strain isolated from Florida and then compare it with other strains." This would be the first time anyone tried to compare the entire genomes of two or more strains of a bacterial species. TIGR's grant called for a novel combination of two rapidly developing scientific fields: comparative genomics and forensic microbiology.

The day before the grant came through, Americans received more shocking news. During a press conference, Homeland Security director

Tom Ridge revealed—perhaps inadvertently—that the anthrax in the letters came from a strain called Ames, which meant the anthrax powder had not been developed by al-Qaeda. It had most likely originated in the United States—perhaps in a military research laboratory. The FBI had known this since October 5, after Stevens was hospitalized but before his death.

When the CDC told the FBI about a case of anthrax in Florida, FBI agents had immediately obtained a sample from Stevens's spinal tap. Agents then asked Paul L. Jackson, a Los Alamos National Laboratory authority on the genetics of *B. anthracis*, what to do next.

"Find out which strain it is," Jackson replied. "Suppose somebody murders somebody, and I say, 'A human did it.'" Saying Stevens was dying of anthrax was just as vague.

Jackson recalls getting "some pushback," but the FBI soon followed his advice and raced samples of Stevens's spinal fluid to two leading anthrax experts: Paul Keim of Northern Arizona University and Tanja Popovic at the CDC in Atlanta.

Early the next day, Keim and Popovic reported disturbing news: Stevens's illness seemed to have been caused by a strain called Ames that had been developed in an army research laboratory in Maryland.

"Oh shit," Jackson said.

• •

When Director Ridge broke the news on October 25, bacteriologists knew exactly where Ames had come from.

In 1981, a fourteen-month-old heifer in South Texas contracted anthrax and subsequently died. A bacterial culture from its organs was sent to the United States Army Medical Research Institute of Infectious Diseases (USAMRIID) at Fort Detrick in Frederick, Maryland. Scientists there determined that the heifer's anthrax strain was especially potent, ideal for testing the efficacy of anthrax vaccines, including one that would be given to American soldiers during the first Gulf War. Because the sample from the heifer arrived in a box marked "Ames, Iowa," the name "Ames" stuck.

The news should have relieved me. As Claire Fraser noted, "The fact that the strain came from a laboratory meant that we wouldn't have to scour the world for shovelfuls of contaminated dirt," the natural habitat of *B. anthracis*, in search of other strains to compare with ours. On the other hand, Ames made Stevens's death look even worse than murder—and more like a national security crisis. Laboratories all over the world—quite literally "everywhere," one expert said—kept samples of *B. anthracis* Ames for research.

Ames definitely posed a big scientific problem. Back in 2001, no one—and I mean no one—could tell which laboratory the Ames spores that had killed Stevens came from. The consensus among scientists was that every sample of Ames looked and behaved precisely the same; nothing a laboratory did could alter a strain of anthrax. Whoever was mailing Ames around must have thought he had the perfect, untraceable murder weapon.

I knew we had to find a way to differentiate the Ames samples held in various laboratories. This was beyond finding a needle in a haystack. We would have to create new science on the fly. It was not going to be easy.

• •

In theory, when a bacterium reproduces, it creates two identical copies of itself. In reality, natural variations can occur. Keeping a bacterium alive for months or years in a laboratory allows it to replicate over and over, giving it many opportunities to make mistakes—that is, to mutate. As a graduate student working with bacterial cultures for my doctoral thesis some forty years earlier, I'd seen bacteria change their shape or appearance, or lose or gain the ability to ferment lactose, or alter other metabolic properties after months and years of laboratory culture. After dealing with scores of bacterial species and strains, I could see a clear pattern: each permanent change was accompanied by a change in the organism's DNA. That meant that if we didn't fully sequence Stevens's anthrax immediately, we might end up sequencing laboratory-induced changes, not the original strain. Fraser had to start working on the Ames samples fast.

Soon after the letters to Senators Leahy and Daschle and the *New York Post* surfaced, FBI director Robert S. Mueller III asked me to come to his office. He wanted my opinion of high-resolution images of the powder in the letters. All the powder, whether sent to New York, Washington, or Florida, had tested positive for anthrax. All derived from the Ames strain. I wasn't told where the samples came from, but it was obvious the powder must have been made in two separate batches; my early work at Georgetown University with electron microscopy expert George Chapman told me that. The spores in the first batch—which I later learned were from envelopes sent to Florida and New York—had a lot of extraneous matter stuck to them. The spores in the second batch—those that were mailed elsewhere in New York and to Washington, DC, and were infecting more people—were clean, sharp, light, and fluffy, and apparently exceptionally pure and powerful. Such refined spores suggested that the person who had loaded the envelopes had both laboratory experience and the type of equipment found only in research institutions run by universities, industry, or the military.

FBI research soon determined that the Ames strain had not been sent to research facilities "everywhere," as we'd been told. Ames had gone to laboratories in only three foreign countries: England, Canada, and Sweden, making al-Qaeda an even less likely culprit. On February 9, the FBI put out a call to the public for information. Their profile pegged the killer as a loner, a single adult male with access to *B. anthracis* and the knowledge, expertise, and laboratory for refining it. Everyone with even a passing knowledge of Ames was a suspect. When Dr. Michael R. Kuhlman at Battelle Memorial Institute in Columbus, Ohio, offered to give one FBI agent a technical tutorial on the aerosol properties of the powder, an agent politely declined. "You don't seem to understand," the agent told Kuhlman, "you're all suspects right now."

But unbeknownst to us, we had already gotten our first break.

• •

Two weeks after Stevens's death, Teresa "Terry" G. Abshire, a highly skilled civilian laboratory technician at the USAMRIID who'd already seen, in her words, a "kazillion" Ames cultures, was growing cells from the powder found in the envelope addressed to Tom Brokaw. She had placed some spores on nutrient-rich agar-agar in a dozen petri dishes, which she kept at temperatures similar to those inside a living animal. These conditions would allow the bacteria to break out of their protective coating and emerge ready to divide and multiply. Within twenty-four hours, Abshire would normally see a raised, grayish-white colony of *B. anthracis* cells on the agar in petri dishes. But Abshire was so overwhelmed with samples to be analyzed that she didn't have time to look at the Brokaw material until it had been growing for forty-eight hours, twice as long as usual. The vegetative cells would have reached a later stage in their growth cycle and taken on a different shape and texture, but this batch didn't look like it was supposed to. Abshire thought one colony looked particularly odd: even without a magnifying lens, she could tell it was more than the usual grayish white.

Wondering if she was seeing a contaminant, Abshire did a series of tests to confirm whether the strange tannish mound of cells was 100 percent *B. anthracis*. It was. Abshire's supervisor suggested she talk with Patricia L. Worsham, a PhD microbiologist who'd studied the formation of irregular colonies created by germinating spores of *B. anthracis*. The women secured more powder from the Brokaw sample for testing, and sure enough, whenever this powder was cultured for forty-eight hours or more, odd-shaped tannish colonies appeared.

When Worsham and Abshire cultivated spores from the letters sent to Senators Daschle and Leahy and the *New York Post*, they found the same odd variant. Then they found more variant colonies with different characteristics. This was painstakingly difficult work, and no one should underestimate the dedication and intelligence of these two highly trained women. They found and saved the variants immediately, before the powders had been subjected to laboratory handling. And because Worsham and Abshire spotted the oddities so quickly, they knew the anthrax powder probably had these characteristics when the letters were mailed. These were the first differences anyone had found between the original Texas cow's *B. anthracis* and the bacterium that killed Stevens.

Worsham took the odd colonies of cells, purified them, and transferred them over and over again to fresh agar plates. Sure enough, the differences in colony shape and color held true. By late November, less than two months after Stevens's death, investigators knew that different laboratory samples of Ames could be distinguished from one another.

This was our smoking gun. But we knew court testimony by two investigators eyeballing "tannish" versus "grayish-white" bacteria would never hold up under cross-examination. We had to tie these visible changes to specific genetic alterations.

• •

Bad news kept pouring in. By November 21, four more people had died: two postal workers, a hospital employee in New York, and a ninety-four-year-old woman in Connecticut whose mailbox had been contaminated with anthrax from an unknown source. (We later learned that the postal service's high-speed letter-sorting machines pressed down so hard that powder could escape through an envelope's pores and into the air, potentially infecting other mail.) All the powder, whether sent to New York, Washington, Connecticut, or Florida, tested positive for anthrax. Recipients in the media and Congress attracted most of the publicity, but the majority of inhalational patients were mail handlers who'd been exposed at work. Thanks to heightened awareness, antibiotics, and aggressive treatment in intensive care units, six of the eleven inhalational anthrax victims survived. Eleven more people developed cutaneous anthrax infections; they, too, survived.

From the time Stevens was diagnosed with anthrax to late November, approximately ten thousand people took antibiotics for possible anthrax exposure, and entire office buildings were closed for decontamination. The post office was sanitizing small quantities of mail with radiation, based on a protocol quickly developed by an interagency group organized by the White House Office of Science and Technology Policy under director Dr. Jack Marburger. As rumors, hoaxes, threats, and panic circled the world, Washingtonians organized anthrax-safe rooms and donned latex gloves and face masks to open their mail. "We were panicked," wrote *Washington Post* columnist Richard Cohen.

In a crisis like this, you might think there would be government procedures for bringing together the best scientific minds as quickly as possible to work toward a solution. But there weren't. There was, however, a growing chorus of government biologists familiar with the

Human Genome Project who believed we should fully sequence *all* the most dangerous microbial pathogens, not just anthrax, to prepare for the possibility of future bioterrorism. The interagency committees I'd worked on had been urging the government to do so.

Sequencing the DNA of living organisms falls under the aegis of the NIH, which had already made a significant investment in sequencing the human genome and infectious pathogen genomes, as well as planning to sequence bioterrorism pathogens, including a number of anthrax strains. Which was why, shortly after Stevens's death, I phoned Dr. Anthony Fauci, director of the NIH's National Institute of Allergy and Infectious Diseases (NIAID). We agreed that we needed to sequence the anthrax isolated from Stevens as soon as possible.

A recently disbanded White House task force with sequencing expertise hastily reassembled. Once it was clear the task force would focus on sequencing bacterial pathogens, I rejoined the group enthusiastically. The group's first meeting was set for the evening of December 18. Generally, interagency meetings must be approved by the Office of Science and Technology Policy. But Ari Patrinos, a leader in the Department of Energy where the Human Genome Project had been launched, knew time was critical. Patrinos is one of those wonderful people who makes things happen, so he ignored regulations and invited relevant individuals from the NIH, NSF, and Department of Energy to the meeting without going through the usual channels.

We hoped our anthrax task force could marshal the nation's resources to find the source of the variant anthrax, but we also had bigger plans. We wanted to prepare the country for the possibility of a dangerous epidemic or a future biological attack. During that first meeting, in Fauci's conference room, we quickly reached a consensus: the country needed a database of genetic information on pathogenic microorganisms. Every nucleotide on a strand of DNA would have to be sequenced; relying on a few sample regions would not suffice. Without complete information, health officials would lose valuable time trying to pinpoint the biological causes of epidemics, attacks, accidents, crimes, and hoaxes. Besides, only complete sequencing could produce ironclad identifications that would stand up in a court of law.

Patrinos was pleased. During intense meetings, he often kept two sets of notes: one in English that nearby snoopers could read over his shoulder, and the other a secret diary in his native Greek for his eyes alone. After the meeting, Patrinos wrote—in Greek—that he was very happy with its outcome. Calling the meeting without consulting higher-ups had been worth the risk.

Then nothing happened.

I waited—impatiently—for a week or so. Time was slipping by, time that we could have spent analyzing the DNA of the anthrax that killed Stevens.

I phoned Tony Fauci again.

"Look, I'll run this thing," I said.

"Fine," Fauci answered. He delegated Maria Giovanni, a talented genomics scientist at NIAID, as his liaison; Jack Marburger soon after assigned Rachel Levinson as the liaison with the White House. (Astute readers will notice the number of women scientists who played leading roles in our endeavor.) As soon as I hung up the phone, the helpless feeling I'd had ever since 9/11 disappeared. I had a job to do. I knew what had to be done, and I had the wherewithal to act. Finally, I could do something constructive.

I needed an interagency team of experts who had power of the purse *and* "hands-on" knowledge of the human genome sequencing effort. They had to be in high enough government posts to speak for their agencies—but not too high. What did most agency chiefs know about sequencing? Not much.

I soon realized the team could not be an official committee. That would have made us subject to the Freedom of Information Act. We would have to remain an informal group of people dedicated to working quietly, confidentially, and unofficially. The chiefs of more than seventeen agencies would know why their genomic experts disappeared for an hour every Friday afternoon, but there'd be no record of our group anywhere. We kept no minutes, held no public hearings, and existed in a strange bureaucratic limbo, known by scores of people but off the books. Despite the fact that we were "just" a group of likeminded people who got together once a week, we called ourselves NIGSCC—the National Interagency Genome Science Coordinating Committee. The NIGSCC would meet for one hour every Friday for three years, beginning in 2002; every other Friday for four more; and as needed for three additional years. People spoke when they had something to say, and at the end of the hour, everyone was gone, action plans in hand.

At our first session, early in 2002, a gentleman at the entrance to the meeting room (known as a SCIF, a sensitive compartment information facility) took our BlackBerries, but inside, the place was like any ordinary conference room: stuffy and filled with too many people and not enough chairs. On an easel, I outlined our long-term goal: identifying dangerous bacteria and viruses. We'd be marshalling the nation's resources to do three things at once: find the source of Stevens's anthrax, prepare for a dangerous epidemic, and provide the information needed in case of a biological attack. The purpose of that first meeting was to match the most dangerous pathogenic organisms with agencies willing and able to fund their sequencing.

I made a list of the most dangerous microbial pathogens: Beth George from Homeland Security said her agency could fund the sequencing of anthrax; she later arranged additional funding for sequencing Ebola and smallpox. Maria Giovanni at the NIH offered to fund sequencing of additional anthrax strains. "NIAID can collaborate with other government agencies like NSF and DOE," she said. And since NIAID had sequencing capacity and data analysis platforms already in place, NIAID could sequence additional anthrax strains beyond the ones it had just started to sequence, as well as other bioterrorism pathogens. "Here's funding my agency could contribute," those with smaller budgets volunteered. Each agency provided funding to outside researchers in accordance with its own procedures.

Our top priority, in those first three years, was anthrax. We needed to prove that when a laboratory grew and experimented on a sample of Ames, the continuous reculturing of the sample would produce variants in it. Some traditional microbiologists may have thought we were trying to do the impossible. When USAMRIID's own anthrax vaccine specialist, Bruce E. Ivins, learned about Abshire's first variant, he dismissed her discovery. He didn't believe a visible oddity could be traced back to a specific genetic change. "You can't really tell one particular strain of Ames from another," Ivins told his boss. At this point in our investigation, he was absolutely correct. One doubting "expert" said we were trying to do "*Star Wars* stuff." Even so, we had to try.

But what strains should we compare? Initially, the FBI wanted Claire Fraser and Timothy Read at TIGR to compare the Ames strain that had killed Stevens with an Ames strain that TIGR was already sequencing for the Department of Defense. The latter strain had come from a British military research laboratory at Porton Down.

Bruce Budowle, the FBI's leading DNA expert and the person who, in my opinion, did most to modernize the FBI's approach to science, took time off from identifying victims of 9/11 to help. Budowle soon realized Porton Down's sample was not an appropriate basis for a comparison. The British had used extreme heat, powerful antibiotics, and harsh chemicals to remove *B. anthracis*'s toxin genes. Those treatments had made it safer for laboratory study, but they'd also induced distinctive mutations; Porton Down's Ames was no longer a close stand-in for the Texas heifer's anthrax. At a meeting with the FBI that both Budowle and I attended, Budowle told his bosses at the agency the bad news: we had to go back and compare the sample taken from Stevens with the original one from the Texas heifer. Fortunately, the US Army's Dugway Proving Ground in Utah had kept a sample of the heifer's original *B. anthracis* in cold storage ever since the animal's death in 1981. That was the virgin Ames, one that had never been altered in a laboratory. From then on, our baseline strain would be the original Texas heifer's Ames, not the ones from Porton Down or Stevens. Claire Fraser's team at TIGR would have to begin anew.

But not everyone was certain that the visible changes Abshire and Worsham had seen were the result of altered DNA. I was confident they were; my PhD thesis work had shown me that growing bacteria on different nutrients over long periods of time could cause such changes in morphology and metabolism. Study of the *B. anthracis* that had sickened and killed US Postal Service employees from Washington, DC, showed that only some of the cells grown from the bacteria were distinctive; but those particular cells retained their distinctiveness even after being isolated and recultured in the laboratory. This suggested that a subset of the cells in the USPS cultures carried a mutation.

To find mutations, TIGR first had to sequence the entire genome of the anthrax bacterium taken from our reference sample (the original strain from the Texas heifer), then sequence the genome of the anthrax powder slipped into the letters, then determine where the two differed, and finally spot those same differences in one or more of the 1,070 samples of Ames strains collected by the FBI from laboratories in the US, England, Sweden, and Canada. It was going to be tough.

We would have to develop new analytical techniques at the same time that we were doing detective work; our success was by no means guaranteed. I felt as if we were designing and building an airplane while we were flying it, and the work wasn't going fast enough. And all the while, we worried the killer might strike again.

I couldn't confide in Jack, who knew I was helping with the anthrax investigation but told me he didn't want to know anything about it—which was good, as I couldn't have told him anyway. We were working on something of tremendous national importance, and while it was frustrating how slowly the work seemed to be moving, we all knew everything had to be done carefully and precisely. Too much was at stake.

• •

An assembly line was formed. First, technicians in Paul Keim's laboratory at Northern Arizona University in Flagstaff used standard bacterial tests to authenticate each sample as *B. anthracis* Ames. Next they purified the DNA by using heat, enzymes, and other chemicals to dissolve the bacterial cell walls and free, clean, and settle out the DNA. Then this DNA was forwarded to TIGR and other laboratories, where scientists used semiautomated techniques recently pioneered for the Human Genome Project to analyze DNA. The DNA was sheared into random lengths and sorted by size before the pieces were cloned and reassembled into their most probable sequence. Eventually, twenty-nine government, university, and commercial laboratories were involved in the investigation, analyzing spore powders and samples from envelopes and post office equipment. The groups worked in total secrecy.

My former student Jacques Ravel joined Claire Fraser and Timothy Read's lab at TIGR in 2002 and became lab leader when Read left to take a position elsewhere. Computational biologist Steven L. Salzberg led Mihai Pop and Adam Phillippy in bioinformatics analysis to interpret the laboratory results.

Sequencing the entire genomes of multiple strains of a bacterium involved new technology that generated new kinds of data and new computational algorithms. Some scientists believed the computational methods and sequencing machines produced errors. Other microbiologists thought that if genetic variants did appear, they were one-time flukes that might revert back in a few weeks. To reduce the number of machine-introduced sequencing errors, Salzberg and his team wrote algorithms to eliminate the least-probable compositions within a sequence. Then the most likely regions of the sequence were sequenced over and over in hopes of finding more mutations.

Meanwhile, Mihai Pop worked on algorithms to assemble the various pieces of DNA into the full genome of the attack anthrax. But when the data appeared on his desk one day, Pop thought his program had made a mistake: it had put together two DNA sequences that should have appeared in different locations. "No one should believe software," Pop says. "We needed to figure out if this was a true biological signal or a software error." The bioinformatics team soon learned that TIGR's lab had taken a shortcut. "We didn't have a lot of DNA available," Pop recalled. "And at every step in the laboratory, you lose some DNA, so the lab had decided to save DNA by skipping one of the steps." That's when the TIGR team realized that the assembler had missed a significant mutation—a large, 1,000-base-pair duplication of DNA—in the attack strain.

• •

The work was extraordinarily exacting—and slow. Ravel and a few colleagues at TIGR worked on a need-to-know basis with the FBI, using state-of-the-art technology and bioinformatics platforms supported by the NIH. TIGR scientists attended so many conferences hosted by FBI agents that Pop almost got used to speaking to audiences full of people carrying guns. But progress was being made. Deep into the project, after nearly a year of work, Ravel and others at TIGR began to find a growing number of telltale DNA signatures associated with Abshire and Worsham's odd-looking variant colonies.

Ravel and several FBI agents with PhDs in biological sciences chose four of the most distinctive mutations to focus on. Following FBI regulations, the samples were secretly coded so only two agents knew where the samples had come from; Ravel and others at TIGR never did. For the next five years, Ravel would search for these four mutations in the 1,070 Ames samples gathered from laboratories in the US, England, Canada, and Sweden. He was trying to find a sample with all four mutations.

In September 2007—nearly six years after the anthrax attack—Ravel was reporting his latest results to several FBI agents when one of them asked, "What about sample so-and-so? Can you show it to me?" Ravel knew the sample had already tested positive for three mutations—and just that month, he had found it tested positive for a fourth. When the agent saw the new result, Ravel watched his expression change from neutral to extraordinarily happy. "I could see that particular sample meant something to him," Ravel would later say.

As a Frenchman, Ravel would normally have celebrated a successful result with champagne for everyone in his lab. But he couldn't tell anyone. He couldn't even tell his lab an agent had smiled at seeing the samples; their work would not be declassified for another year. "I did, however, feel that five years of work had contributed something," Ravel said later. "Sometimes in science, you publish some work that interests only a few people in the world. But this time I felt that we had a much larger impact."

Our genomics hunt had finally produced the vital evidence: altogether, four genetic changes had been identified as the molecular fingerprint of the mailed spores. And eight of the 1,070 Ames samples collected by the FBI had all four of those changes. Armed with that knowledge, the FBI could use traditional detective work to search for the person or persons who had mailed the letters containing anthrax spores with those four mutations.

And soon, the FBI could say that the only Ames samples testing positive for all four genetic changes came from one particular flask at USAMRIID, where Worsham and Abshire had identified the first mutations. It was Flask RMR-1029, secured by magnetic card access and stored in anthrax vaccine expert Bruce Ivins's walk-in refrigerator in the biocontainment laboratory where he worked. By mid-2008, the FBI had established to its satisfaction that Ivins was the only person who had ever worked alone around anthrax samples from that flask, thus clearing a longtime FBI suspect who had never had access to the flask.

We will never know for certain what exactly happened and whether Ivins was working solo or with others—or whether he was truly even involved—because, as the FBI was preparing to arrest him, he took enough Tylenol to put himself in a coma. His death three days later on July 29, 2008, was ruled a suicide.

Ivins had codeveloped an anthrax vaccine, which was administered to Gulf War veterans. Recipients of the vaccine had complained that it caused a cluster of chronic symptoms, including fatigue, joint pain, and headaches. Many people thought that if Ivins did indeed send the letters, he was likely motivated by fear that his vaccine program was in danger of being shut down, and believed that panic over the anthrax letters would prove it was still needed. But his death makes Ivins's motivations and intentions impossible to determine—much less understand. On August 18, 2008, the FBI held a press conference to explain the science behind the investigation, the largest ever in FBI history. The FBI did almost all the talking; I was introduced as a former director of the National Science Foundation, "which provided funding for much of the genetic sequencing effort," and spoke only briefly, mentioning the consortium of agencies that had worked together to create "a legitimate new discipline, namely, microbial forensics."

After Ivins's death, the FBI declassified the investigation and we got a possible answer to something that had puzzled us for seven years. Why, when the DNA of *B. anthracis* changes so rarely, did the spores in the letters contain so many altered genes? The answer is probably that Ivins's legitimate research involved building anthrax vaccines by combining Ames material from different laboratories. By collecting different anthrax strains, he was also collecting the different stress-induced mutations that manipulations in those laboratories had produced. Growing quantities of Ames had also, apparently, made his mixture acquire two more variations of its own, purely by accident.

Much later, I was given access to a redacted analysis of Ivins's psychiatric condition, made after his suicide. The report, issued "without dissent" by a nine-member independent panel, cites "a significant and lengthy history of psychological disturbance and diagnosable mental illness . . . that would have disqualified him from a secret level security clearance had they been known." After Ivins's attorneys had him he would probably be indicted, he attended a group therapy session in which he blurted out a plan to murder coworkers. He was quickly and involuntarily committed to a local hospital and then to a mental hospital for more than a week. The psychiatric panel concluded that hospitalizing Ivins "likely prevented a mass shooting and fulfillment of his promise to go out in a 'blaze of glory.' "

Two years after Ivins's death, the FBI published the 92-page *Amerithrax Investigative Summary*, recapping the seven-year-long investigation. Its first page credits "groundbreaking scientific analysis that was developed specifically for the case" by persons or organizations unnamed that did the work of tracing the anthrax powders used in the attacks back to Ivins's flask.

• •

The NIGSCC's scientific investigation lasted seven years, much longer than we expected, and I continued to serve as chair until the group disbanded. We were naive in the beginning. There was so little sequencing data for pathogens that we thought sequencing three strains of a species, maybe even five strains, would tell us enough about the species to figure out where the attack anthrax had come from. But anthrax mutates so rarely that we had to sequence many more strains in order to find differences. The analyses were tedious, severely time-consuming, and expensive.

Today, whole-genome sequencing methods have been employed to fight numerous infectious diseases, including SARS, SARS-CoV-2, Listeriosis, *Streptococcus*, MRSA, swine flu (H1N1), *Klebsiella pneumoniae*, Ebola, and Zika. Sequencing is now widely used to track outbreaks and is the basis of the emerging science of precision medicine. Ravel and his TIGR colleagues would use a similar approach to identify ricin, a dangerous protein found in castor seeds and mailed in threatening letters to Michael Bloomberg when he was mayor of New York City.

For many of its participants, the NIGSCC was the high point of their careers, and several members were decorated for their service. The CIA awarded medals to a number of us, including John Phillips and his team of experts, Ronald A. Walters (NIGSCC's extremely efficient executive secretary), and to me. I continued to chair the NIGSCC until it disbanded in 2011.

The NIGSCC was a remarkable example of what extraordinary teamwork can do; Claire Fraser called us a "spectacular orchestra." Coordination is key to dealing with any crisis. We worked together rationally and fairly. No one gave dictatorial orders. Gender was never discussed, never an issue. We were too focused on what we could do to solve a terrible problem.

The NIGSCC investigation was also proof that no single agency, research institution, or industry has all the expertise needed to respond to every conceivable emergency. And yet because of retirements, budget cuts, and political constraints, assembling a formidable scientific team of federal government employees may no longer be possible.

Looking back on this experience, what pleases me most is that fundamental scientific inquiries into human genetics helped unravel the mystery of the lethal anthrax powder. However, the investigation was also immensely intellectually rewarding for me personally.

By the end of the anthrax investigation, I was keenly aware that the international scientific community needed a rapid and accurate way to identify dangerous pathogens, and I decided it was time to invent one. I would turn to private enterprise for funding and support. There I would find a whole new set of opportunities and—no surprise—discover a whole new set of problems for women in science.

chapter eight

From Old Boys' Clubs to Young Boys' Clubs to Philanthropists

The anthrax investigation was indeed a life changer. It had taken six years to identify the murder weapon, during which time innocent people were killed, hurt, and falsely accused of murder. If we'd been able to identify the dangerous microorganism at play faster, and accurately, we might have been able to prevent some of those tragedies.

I had been identifying microorganisms since graduate school, so I began devising a way to use DNA sequences to identify—within minutes—all the microbial pathogens in almost any kind of sample. It wouldn't matter whether the sample was from Chesapeake Bay water or domestic sewage, a patient's rectal swab or blood draw, ground soil or debris collected from the air. DNA analysis could be used to determine the presence and relative abundance of bacteria, viruses, parasites, and/or fungi, and their species, strains, substrains, and characteristics. My idea involved using computers, genomics, and probability mathematics to match the sequenced pathogens in the samples to a library of data. I believed the method could save many lives—and revolutionize microbiology. That was my dream.

In 2004, a few months before the end of my sixth and last year as the NSF's director, I resigned and spent several months turning the idea into a data management system. Developing the algorithms into a workable method would have cost more than I'd receive in an NSF or NIH grant, however. I decided to try my luck in a field I had never thought of entering: the world of business and entrepreneurship.

Over the next ten years, I wound up trekking from one part of the business world to another, from a huge multinational conglomerate to my own company to nonprofits. I would discover business to be even more misogynistic than universities and government, but I would also find female venture capitalist companies dedicated to supporting women entrepreneurs and congenial nonprofits where a scientist could tackle problems of global significance. Those years were a vast learning experience—one that I hope sharing will help the many scientists who are moving beyond academia and government today.

Drastic cutbacks in federal and state funding have turned scientific research into a high-risk career. Half of our PhD scientists are already migrating out of academia, and in 2017 only one in four or five life science PhDs had tenure at a university or a position that could lead to tenure. PhD scientists in other fields are leaving government as well. Young and old, scientists who need to fund their laboratories and their students, or who themselves need jobs, are going hat in hand to private enterprise, venture capitalists, foundations, and the military to seek financing. These are all male-dominated worlds where women usually lack connections. Professors are now judged not only on their research productivity, publications, and teaching but also on the number of patents and board seats they hold and the start-ups with which they're involved. Because of this, even senior professors must learn how to work with business.

I disagree with those who say this country is producing too many science PhDs. To keep growing, our highly technical economy needs them all—and more. To succeed in science today, however, every person must plan for his or her future. And for many, that future just might involve a sojourn in business or industry.

• •

I was unprepared for what lay ahead. Like many women scientists moving into business, I joined an established corporation. Canon, a Japanese multinational conglomerate specializing in optical and imaging products, had asked me to advise it on starting a diagnostics subsidiary as its entry into the life sciences market. The job seemed a natural fit. I could develop my rapid microbial identification method within the subsidiary while continuing to do my academic research and volunteer in scientific organizations. So I began a new career as chair and senior advisor of Canon U.S. Life Sciences, Inc.

It was a great job title. But setting up what was to be a subsidiary of a multinational conglomerate taught me that large corporations can be even more bureaucratic and autocratic than academia or government. Not to mention that as the prestigious nonprofit Catalyst warned, gender-based inequality was entrenched in the corporate world.

Even today, women graduates of elite MBA schools lag behind their male classmates in pay and position. As late as March 2019, only twenty-four women were CEOs of the companies that make up the S&P 500. Suspicions are growing that when women are appointed to top jobs, it's because their companies have serious problems that men don't want to tackle. Christine Lagarde, the first woman president of the European Central Bank, calls it a "glass cliff": when the woman fails, the men around her are absolved of responsibility.

In particular, women have not done well as senior executives in Japanese companies. Indeed, when I served on an advisory panel for a Japanese government agency in the early 1990s and was touring a major laboratory in Japan, I met a woman with a PhD who was working as a low-level laboratory technician; I was told this was customary. Today the Japanese government is consciously trying to increase the number of professional women in professional leadership positions. Canon, like most highly successful Japanese companies, is trying, too, but having a difficult time moving away from traditional practices and attitudes.

As it turned out, developing a new, untried idea for medical diagnostics was considered too risky, and after three years at Canon, it was clear the company preferred to follow a path more closely aligned with its strength in imaging and camera manufacturing. It was the right time to move on to new challenges. I had never thought of becoming an entrepreneur, but I needed to be my own boss, and—as I'd learned from my years in universities and government—controlling the money is critical. I would have to form my own company. I thought it shouldn't be so hard. I'd directed the NSF, a multibillion-dollar agency, and during my tenure, the NSF had been awarded a crystal eagle from the US Office of Management and Budget for being the best-managed government agency in the country. Besides, I'd served on plenty of corporate boards. So in 2007, I boldly started my own company, CosmosID. My goal was not to become a billionaire. I just needed to raise enough money to keep developing methods for modernizing microbiology and diagnostics and improving medical care.

••

I had not realized how fundamentally different business and academia are—from their ethics and goals to the people and data you could trust to what could and could not be published to whose orders were above question. Ostensibly, most scientists share the same goal: to unveil the magnificent laws of nature. Business has rules, too, but its ethics vary according to the boss, the company, and the industry. In a university, intellect is the ideal measure of a man or woman; in business, your worth is measured by the money you generate for the company.

Many of the groups that support women in business today did not yet exist or were just starting out in 2007, and I didn't know where to turn for advice. I was invited to speak to an organization for female biologists interested in business, but in talking with them, I realized they needed help with basic business strategy. I'd already assembled a formidably knowledgeable advisory board of investors from industry, academia, government, and the financial world. But sadly, Robert Porter and Rod Frates, two initial investors and advisors who could have helped guide me through unfamiliar business terrains, died soon after the company was formed. They were excellent businessmen with kind hearts and had been hugely supportive. Without them, I had no one from whom to seek advice on critical issues like whether to retain two lawyers: one for the company and another to represent my own interests, especially after new investors were brought in and I was no longer the major investor. And so, as a rookie bereft of mentoring, I made three serious mistakes that made the first nine years of CosmosID more difficult than they needed to be.

First, I made strategic errors. The company was launched with $2 million from key investors and $2 million in grants, the largest from the Department of Homeland Security to identify dangerous microorganisms rapidly and accurately. The company had an executive officer, a chief scientist, and several staff members, mainly computational scientists and engineers. But the first CEO believed producing revenue immediately should be the top priority, and so the company burned through that early money working on product development too soon, when the focus should have been on financing further development, publishing our findings and early successes in detecting pathogens, and building a reputation for the company. A product launch would then have had a foundation on which to build.

Second, identifying a competent business executive is critical, but finding a CEO who also understood enough relevant science to operate a biotechnology firm was difficult, at least in those early days. Research scientists at Yale, Columbia, and the University of California Santa Barbara have found that men overestimate their abilities by as much as 30 percent, although this research, unfortunately, did not appear in time to guide me in choosing the company executive. "The men go into everything just assuming that they're awesome and thinking, '*Who wouldn't want me?*' " Victoria Brescoll of the Yale School of Management told *The Atlantic*. Ernesto Reuben of Columbia Business School concluded that men's overconfidence in their abilities helps explain why fewer women make it to the top leadership ranks. Disappointingly, some researchers don't believe the solution is for men to estimate their worth more accurately; they say women should behave more like overinflated males.

Another big reason for my hiring mistakes was that scientists generally trust their colleagues. After all, their publications are reviewed by peers and appear in journals for other experts to critique, confirm, or debunk. Business has no such peer-review process. In science, your work has to be accurate. In business, profitable is good enough. So it was deceptively easy to accept references at face value. Today I know better, and check references, references of references, and *their* references.

Third, and most important, my philosophy was that the company would be a team. That was how I'd operated in science and government—and in fact, the CosmosID scientific staff functioned and still functions as an inspired and inspiring team. But, in the spirit of camaraderie, I shared ownership in the company with key managers early on, assuming their commitment to the company was as intense as mine.

When problems arise internally, knowing how to negotiate from a position of strength is very important. I'd negotiated successfully for organizations, but—as Linda Babcock and Sara Laschever point out in their 2003 book *Women Don't Ask*—even women who negotiate skillfully with labor unions and donors cannot do the same for themselves. Babcock says women are educated to be caretakers and advocates for others, but when we take the reins, we're branded as bossy and difficult.

What really disappointed me about business was its Wild West practice of rustling ideas. Familiarizing myself with a well-known foundation's grant application form, I discovered that the foundation insisted on owning all intellectual property developed with its funds. Why should ideas and discoveries go to someone who might not use them, or who might bury them if they compete with another product the organization owns? Ideas should be shared and used to help people, especially those in greatest need. I was beginning to learn that I was a scientist, not a profit seeker.

Then I read Katty Kay and Claire Shipman's article "The Confidence Gap" in *The Atlantic*. The article discusses how confidence is as important in business as competence—and yet women too often lack the self-assurance they need to succeed. It made me realize that, given my fifty years of experience, I was a member of a select group of people who knew a lot about identifying microbial pathogens—and I knew we really needed to modernize the way we made those identifications. *Dammit*, I told myself, *this company was my idea, and it is going to succeed*. And as I write, a dozen years after the formation of CosmosID, the company has launched several medical and food and water safety applications, and its future is promising.

• •

Should women scientists steer clear of business? No, definitely not.

I believe that attitudes about women—and minorities—in business are at a tipping point, because executives are learning that we can help expand the economy. Cultural change requires decades. Transforming society's fundamental views about women won't happen for many more years, although such change is absolutely necessary. However, for perhaps the first time in history, powerful men are recognizing their companies will make more money if they give women a share in leadership. Financial services giants like Credit Suisse, McKinsey, Bloomberg, Ernst & Young, and BofA Securities are spreading the message. An International Monetary Fund (IMF) study of two million companies in Europe found that adding one woman in a senior position to a high-tech manufacturing or knowledge-intensive service company is associated with a 34 to 40 percent return on assets. Why? Because such organizations need the "higher creativity and critical thinking" enhanced by independent views. So should women hold *all* senior positions in the corporate world? No. The IMF paper concluded that the peak optimal share of women in senior positions is about 60 percent. It's not women per se who make profits rise; it's different perspectives that bring in new ideas.

The last S&P 500 company without a female director added one in 2019. But that doesn't mean women are going to achieve parity with men overnight. In fact, the turnover in executive boardrooms is so low that studies suggest women won't reach equal representation with men for forty more years. It also takes time before a board listens to any newcomer, whatever their gender or ethnicity. And new women board members aren't being given the same leadership opportunities as men: women are usually given shorter terms and fewer committee chairmanships. Bring in two or three women, and they can support each other's arguments and make change happen faster. Some European countries—as well as the state of California—are mandating quotas for women on boards of directors. Quotas are a possible solution, but they're contentious because some consider them affirmative action.

• •

Diversifying technology companies will be even more challenging. Approximately 25 percent of the growth in this nation's GDP since the 1960s has come from opening up the fields of law, medicine, science, academia, and management to black and white women and black men, according to the National Bureau of Economic Research. But when the world wide web started gaining steam twenty-five years ago, it was mostly risk-loving young men who jumped aboard. As a result, the revolution in information technology and e-commerce—with profits, glamour, and power—whizzed right by women of all ethnicities and African American and Latino men. In the tech industry today, white and Asian males make up approximately 70 percent of the employees at Microsoft, Google, Apple, Twitter, and Yahoo—while women are often concentrated in lower-status sales and marketing jobs that don't lead to patentable, profitable discoveries.

Worse, more than half the women in science, tech, and engineering abandon their companies when they hit the glass ceiling mid-career. The first two big studies of corporate America and women in STEMM reported the same phenomenon. *The Athena Factor*, by Sylvia Ann Hewlett and colleagues, found that in 2008, 52 percent of mid-level, thirty-five-to-forty-year-old women with degrees and significant experience in science, technology, and engineering planned to leave their jobs—not to start families, but because their prospects for advancement were so slim. "Climbing the Technical Ladder," a 2013 study by Caroline Simard and Andrea Davies Henderson, found that 56 percent of women in tech quit mid-career after reaching the glass ceiling.

If half the *men* in tech left, the country would declare a state of emergency, Hewlett noted. American businesses already complain that they must hire foreign nationals because a shortage of STEMM workers limits their growth. If the attrition rate of female STEMM employees was reduced by 25 percent, Hewlett calculated that companies could add 110,000 highly qualified workers to their payrolls.

The frat-house atmosphere of some Silicon Valley start-ups has driven many women from tech. The early days of Uber, for example, were alcohol-drenched, with open beer kegs on each floor 24/7. When Uber's board hired former US attorney general Eric Holder to investigate complaints of sexual harassment in preparation for the company's initial public offering, Holder recommended a staggering forty-seven different steps to take to clean up the company's image. Employees at other early start-ups have mimed masturbation on an industry conference stage, advertised women as "perks" at company events, and sold apps for Wobble iBoobs, Titfinders, and Titstares. Too many Silicon

Valley leaders are also outspoken misogynists. Peter Thiel, an early investor in PayPal, YouTube, LinkedIn, and Yelp, wrote in 2009 that giving women the vote was bad for democracy and capitalism. As a result of toxic work environments, "women have been systematically excluded from the greatest wealth creation in the history of the world and denied a voice in the rapid remodeling of our global culture," journalist Emily Chang wrote in *Brotopia: Breaking Up the Boys' Club of Silicon Valley.*

In addition, tech's rampant ageism operates "all the time" against senior women, too, says Amy Millman, the founder and president of Springboard Enterprises, a venture capital fund that has raised more than $8 billion for more than seven hundred women founders of tech-oriented companies. "In the old days," Millman says, "you couldn't apply for a senior position if you were under 40. Now it's like, 'Why would you hire anybody over 50? What can they possibly know about what this business needs?'" When one middle-aged woman scientist sought funding from investors, the men—many with gray in their beards—looked at her and said, "What you really need is a twenty-something guy to be CEO."

• •

Women who want to start their own businesses face an especially difficult problem: the all-male venture capital industry. I've spent a lifetime working around barriers and climbing over obstacles, so believe me when I say that for a woman entrepreneur who needs money to start and grow her company, this one is really tough. She'll need a heavy coat of armor.

Venture capitalists perform a public service; they provide funding that new companies need to survive and prosper. However, the venture capital world is composed almost exclusively of rich men. Even though it's been proven that diversity produces better profits, companies with women founders received less than 3 percent of all venture capital funding in 2018. The venture capital industry is "staggeringly" homogeneous, report Harvard Business School professor Paul Gompers and his colleague Silpa Kovvali. And note that the word "staggeringly" is theirs, not mine. They looked at every venture capital organization in the US since 1998, and found that the industry has remained remarkably uniform for twenty-eight years. Only 8 percent of investors are women, 2 percent are Latinx, and fewer than 1 percent are black. One in four venture capitalists with an MBA obtained their degree from the same institution: Harvard Business School. As of 2018, nearly three-quarters of venture capital firms had never hired a woman investor; women could be found as chief financial officers or marketing and communications directors, but not in roles that would have them deciding which companies to fund. Entrepreneurs from one generation of tech companies invest in the next generation of tech companies, thereby preserving and concentrating wealth in one small group, writes historian Margaret O'Mara of the University of Washington. Economists Alison Wood Brooks and Fiona E. Murray discovered that even when male and female entrepreneurs make identical pitches, rich white male venture capitalists prefer to invest in men, preferably attractive ones. The problem is then compounded, because male-led start-ups tend to appoint male executives and male advisory boards. Interestingly, however, if a senior partner has a daughter, his company is 25 percent more likely to hire a female partner.

Carol A. Nacy, a former president of the American Society for Microbiology and a friend, has been mistaken for a secretary four different times while making pitches to male venture capitalists for one of her three companies. According to script, one of the men says something like, "Hon, could you get me a cup of coffee?" and Nacy answers cheerily, "Sure, what'll you take in it?" Then, after serving him coffee, she enjoys watching the guy's face as she's called to the lectern to make her pitch. After years of practice, she's figured out how to game the system. "When I go to a venture capital group of only men," she says, "I do most of the speaking, and my chief business officer and I watch faces. If they're incredulous about something I've said, he'll repeat the same thing—and then it'll be just fine."

The odds may not be in our favor, but clever women scientists are figuring out how to navigate the world of business, just as we've had to for years in academia.

••

I believe universities should do more to help diversify science, engineering, and technology companies. The Bayh-Dole Act of 1980 gave universities the right to own intellectual property derived from faculty members' federally funded research. And today, universities allow investors to use faculty research to start companies with almost all-male management teams, all-male boards of directors, and all-male scientific advisory boards—even in fields where women researchers are leaders. Universities should establish protocols that give investors permission to use faculty members' discoveries only if there's evidence of diversity on their boards—a change that might happen . . . eventually . . . perhaps.

Students are especially vulnerable, of course. One day, a talented female postdoc came into Nancy Hopkins's office at MIT, started to cry, and couldn't stop. She had a problem: every lunchtime, she explained, the male trainees in her PhD advisor's laboratory marched into his office to discuss the companies they were starting; women graduate students and postdoctoral fellows were not included in those conversations and, all too often, were left working in the lab. "We're not giving our students equal opportunities," Nancy Hopkins realized. The MIT Miracle (see chapter 4) hadn't changed her male colleagues' thinking. "They just went out and duplicated the universities' discrimination against women in the venture capital world," she said. She, Susan Hockfield (a former MIT president), and Sangeeta Bhatia (an MIT professor of engineering) found that fewer than 10 percent of the 250 start-ups founded by MIT faculty members had been founded by women, although women make up 22 percent of the faculty. Another study at Stanford found a similar discrepancy: only 11 percent of start-ups by university faculty had a woman founder, although the faculty was 25 percent female. If men and women at MIT started companies at the same rate, we would have forty more biotech start-ups making new discoveries.

While universities have long been far ahead of the business world in supporting women in science, some now depend on generous donations from the superrich. This so-called science philanthropy provides almost 30 percent of the annual research funds at leading universities—enough to influence what gets studied and by whom. Most of these donors are interested in applied, not basic, research—the difference being that basic research into Alzheimer's might try to understand how neurons degenerate with age, while applied research would try to make a drug that reduces amyloid plaque. In any case, more than 57 percent of large gifts from individual donors goes toward biomedical research, and while there are no data to prove this, it's a good guess that most of it probably goes to the nation's top ten schools, certainly to the top fifty. And although some of these ultra-rich donors are advised by boards of scientists, Marc Kastner, former provost of MIT's School of Science and former president of the Science Philanthropy Alliance and now lead advisor to SPA, believes many of them are not—and what's more, they avoid outside peer review for the research they are funding. This worrying trend may be leaving science to the whims of a small number of wealthy individuals untrained in research, MIT economist Fiona Murray warned in 2013.

In any case, between 40 and 70 percent of federal grant money given to universities goes to support the schools' overhead costs for libraries, energy, security, and so on. But donations from scientific foundations and philanthropists generally do not include overhead. If private donations continue to play a major role in scientific research, how are we going to pay for our universities—or for the diversity measures they adopt?

• •

So how can women scientists prepare themselves for the inhospitable corporate world? PhD training can teach women how to be good scientists, but it doesn't teach them how to start or run a business. And getting advanced business training quickly and succinctly is difficult. When I asked one university president, who was my boss at the time, for approval to take advantage of an opportunity to learn the details of financial management, he said no—you learn on the job. Early in my career, I had joined the board of a publicly traded corporation that paid me to attend a two-week Harvard Business School program on the role of corporate board members. The course proved extremely helpful in explaining the duties of a member of the board, but wasn't enough training to start and run a company. I needed to understand the nuances of company leadership and management strategies.

An MBA is seldom the answer. After spending ten years working on original research as a PhD student and a postdoc, few scientists want to spend two more years in a master's program designed for recent college graduates. One solution would be to modernize PhD programs so that graduate students in the sciences who are contemplating business careers can take courses like marketing and business finance or spend a semester as an intern in a biotech company. I suggested this when I served on committees of the National Academy of Sciences and the Association of Public and Land-grant Universities but found that modernizing the PhD was like the idiom "trying to move a graveyard"—extremely difficult if not impossible.

In 2009, I chaired a National Academy of Sciences committee on professionalizing the master of science degree to prepare STEMM undergraduates for alternate careers. There I was able to help lead to the creation of a new kind of master's program for science majors interested in business. Traditionally, master's degrees in science had been regarded as consolation prizes for students who don't pursue a PhD. However, the professional science master's degree—the so-called "science for business" master's—has attracted large numbers of women who've gone on to become science and technology managers, investment analysts, and forensic scientists for criminal justice laboratories.

In these changing times, universities must adjust their PhD programs to train more than academic professors, or their PhD programs may become irrelevant. Most universities now have offices helping faculty members commercialize their discoveries, and students may need their help, too.

Fortunately, some women aren't waiting for academia to change. Twenty years ago, when Springboard Enterprises was started as a nonprofit venture capital fund, few women were interested in investing in women-led companies. Now they are, says founder and president Amy Millman. Springboard organizes training sessions and pitch nights, and provides advisors and templates for start-ups. Millman's dream is to build a parallel universe of women-owned venture capital firms that support and invest in women-led companies. She's not the only one. Candida Brush, a professor of entrepreneurship at Babson College in Massachusetts, is bringing together researchers, educators, and entrepreneurs to find new ways for women to acquire growth capital. Individual entrepreneurs, both men and women, are mentoring female founders now, and my experience with women's networking organizations tells me they can be very effective.

Despite the business world's growing support for women scientists, many still worry that working with corporate entities will compromise their science by limiting the research they can do. I understand that. Despite their concerns, though, two of my favorite activities today are financed by the corporate world. Neither activity was established to make money, but they show that a woman scientist can find deeply satisfying ways to work with business interests.

In my case, each opportunity started with a phone call.

••

On April 20, 2010, three years after I founded my start-up, CosmosID, an oil-drilling rig leased by the London-based international energy conglomerate BP exploded off the southeast coast of Louisiana, killing eleven men. Deepwater Horizon was one of the biggest environmental disasters in US history and the world's largest oil spill into marine waters. BP's reputation took a big hit.

Two weeks into the aftermath of the spill, I received a phone call from physicist Ellen D. Williams, then BP's chief scientific officer. I knew her when she was a professor of physics at the University of Maryland; she is now a distinguished professor and director of the Materials Research Science and Engineering Center at the university. Shortly after the explosion, BP pledged $500 million to study the impact of the oil spill on the Gulf of Mexico's environmental and public health, as well as ways to mitigate future oil spills—because there will be more. (This money was separate from the additional billions of dollars in fines and settlements the company would eventually have to pay.) Williams asked if I would set up and run BP's research program. The half billion dollars in funding would be spent over ten years, with no strings attached.

This was an unprecedented amount of funding for science in the Gulf of Mexico, and there were no existing rules on exactly how it was to be used. But it was an opportunity to create something good out of a disaster. I had experience building scientific organizations from the ground up, and my background as a former director of the National Science Foundation and president of the American Society for

Microbiology, the American Association for the Advancement of Science, and the International Union of Microbiological Societies could lend credibility to BP's plan. In addition, much of my early research had involved oil pollution in marine waters and the microbial biodegradation of hydrocarbons.

Would BP keep its promise to be hands-off? I asked Williams. Scientists would have to devise the research and how and where it was done.

Yes, Williams promised. Absolutely.

If BP really meant what it said, we'd have the freedom to design a new way to fund research. I envisioned using NSF procedures as a template to establish a Gulf of Mexico Research Initiative. Funds would be granted to qualified scientists after an open competition and would be used for research activities such as sampling, modeling, and data analysis; results would be published in peer-reviewed scientific journals; and all the data collected would have to be made available to the public in established databases. All this would be done without interference from BP. When I told Williams what I was envisioning, her response was, "Of course, that's the way it should be."

Such research could address critical societal problems. Runoff from fertilizer and manure, aided by decades of lax law enforcement, had created a vast lifeless zone in the Gulf, but government agencies were providing less than $10 million a year to study the Gulf's ecosystem. I thought that if international experts worked with local scientists, the Gulf of Mexico Research Initiative (which we shortened to GoMRI, pronounced *GOM-ree*) could enhance the scientific research capabilities of the universities in the five Gulf states: Texas, Louisiana, Mississippi, Alabama, and Florida.

If it worked, the initiative would demonstrate a responsible way for philanthropists, politicians, corporations, and venture capitalists to fund top-notch scientific research. It would show that quality science can prevail when industry and private money provide funds and general direction, but scientists decide what's funded and how the research should be done to address societal challenges. As it turned out, it was also a good opportunity to advance women in science.

So I told Williams that if BP really and truly agreed to these operating principles, I'd do it.

We'd have to launch GoMRI quickly. During the four months it took to seal the mile-deep wellhead, Deepwater Horizon spewed an estimated 206 million gallons of crude petroleum into the Gulf, killing marsh grass, birds, fish, and marine mammals, and damaging the Gulf's seafood and tourism industries. To prepare for another oil spill, we needed to know what to do when a spill occurred and how to clean up afterward.

After I signed on as chair of the research initiative and discussed candidates with BP, six world experts I knew and respected were appointed to the GoMRI board. Four were from leading oceanographic research centers: Scripps Institution of Oceanography, the Monterey Bay Aquarium Research Institute, the Woods Hole Oceanographic Institution, and the National Oceanography Centre in England. I asked Margaret Leinen, the director of Scripps, to serve as vice chair. Charles "Chuck" Wilson from the Louisiana Sea Grant College Program would become the chief scientist in charge of the program's day-to-day management. And soon Wilson and I would be communicating by phone or email almost every day, discussing the progress of the research programs or expenditures or any of the other urgent issues that arise when managing a large enterprise.

As early news of BP's half billion dollars spread, Louisiana state senator Mary Landrieu led Gulf-area politicians to the White House to demand control over the funds. The White House, BP, and the five Gulf-state governors were developing a contract to ensure responsible fiscal practices. I worried that if the governors appointed political cronies to our advisory committee, decisions about needed research would be subject to political biases and the money would be spent on big-ticket items like buildings, casinos, and cruise ships. Fortunately, BP was firm. As a political compromise for the first year only, $45 million was divided among the five Gulf states and each governor was given the power to appoint two scientists to our scientific advisory committee. At the same time, as chair of the governing board, I insisted on adding to the contract some important language to keep scientists in control of the science to be done.

It took a lot of fine print. The contract mandated a board of twenty academic scientists that would make all funding and research decisions.

The contract also stipulated that all the scientists on the board would have to have "peer-recognized research credentials and [come] from academic institutions . . . or from other nationally recognized research entities." No members of the research board could be "political appointees, BP employees, or State personnel outside of academic or research institutions." Nor could any members represent any constituency, stakeholder, or interest group. Each governor would choose two board residents from a list of scientific experts residing in their states; the list would be provided by my committee.

Working quickly, the six original board members and I combed through the faculty directories of Gulf-area colleges and universities and identified their leading marine scientists. We gave each governor a list of thoroughly vetted names to choose from. Every scientist invited to join GoMRI's governing board accepted—ten excellent new members—which I took as a sign of the importance of our endeavor.

Within a year of the spill—lightning speed for such a major undertaking—GoMRI-funded scientists were collecting valuable samples of oil and dispersants in the air, coastal marshes, sediment, shallow waters, deep waters, coral reefs, insects, and commercial fisheries. But while we were still deep into our research, the public needed answers. People were asking questions like, "Is the oil going to be here forever?" "Are the fish safe to eat?" "Can I go back to the beach?" "Are my kids going to get cancer?" They needed authoritative, factual answers, so every GoMRI-funded team was required to spend part of its funding on public outreach. A number of excellent women oceanographers working on GoMRI-supported research projects turned out to be superb communicators. Mandy Joye, an oceanography professor at the University of Georgia, often explained Gulf science to the media during the crisis. Years later, people who'd seen her on TV still stop her on the street to ask, "Are you the Gulf lady?"

We contracted with filmmakers to chronicle GoMRI's progress for TV, school, and public use. When the first film showed one white male expert after another, I insisted the next two films provide a balanced and accurate representation of women's contributions to the oceanography work being done in the Gulf of Mexico.

Thanks to GoMRI scientists, we now know more about how to treat oil spills with ecologically friendly dispersants that help clean the oil film that forms on the water's surface and elsewhere. GoMRI scientists identified bacteria in the Gulf that digest spilled oil quite efficiently. In fact, Joye suggested that in any future spill, responders should consider adding nutrients, which would essentially fertilize the water affected by the spill and encourage growth of oil-degrading bacteria. As an added bonus, as Hurricane Isaac charged through the Gulf in 2012, a little flotilla of buoys that we had launched into the Gulf's water system to study currents produced data that, when analyzed, gave us unprecedented pictures of how a hurricane moves through water. First responders now know how the Gulf's surface currents shift under different wind and wave conditions.

As for building a scientific community, we funded a new generation of Gulf scientists: 455 postdoctoral positions, 630 PhD students, 562 students in master's programs, 1,048 undergraduates, and 115 high school students, some of whom will spend their careers in the Gulf area. With the 4,312 personnel members involved in the Gulf of Mexico Research Initiative, we were turning the Gulf into a world-class center for oceanographic studies that will, I hope, show voters how scientific research can benefit an entire region.

• •

GoMRI was just getting under way when I got a phone call from Kurt Soderlund, the founding CEO of Safe Water Network, a nonprofit started by a group of philanthropists that included the Oscar-winning actor-directors Paul Newman and Joanne Woodward, and John C. Whitehead, former chair of Goldman Sachs and one of President Ronald Reagan's deputy secretaries of state. Founded as a nonprofit in 2006, the Safe Water Network aimed to use standard business practices such as realistic pricing to help bring safe water to underdeveloped parts of the world. Soderlund asked if I would consider joining the board. After my years of working on cholera in Bangladesh, I was very interested. Soderlund warned me from the outset, though, that many of the other board members were executives from big American corporations and tended to pride themselves on knowing how to get things done. Some of them suspected that academics—like me—did not.

Sure enough, when I went to meet Safe Water Network's board in New York City, Whitehead put me on the spot. "We're looking forward to meeting you," he said. "We want to know if you're a doer."

"That's good," I responded, "because I'm here to find out if this is a group of doers, and if it's not, I'll be on my way."

The World Health Organization and UNICEF say that one in three people on Earth—some 2.2 billion people—lack safe water. In addition to cholera, more than twenty-five diseases are transmitted via water, including salmonellosis, shigellosis, *Campylobacter*, *Helicobacter*, *Giardia*, *Cryptosporidium*, *Rotavirus*, *Norovirus*, and more. Half the hospital beds in the developing world are occupied by patients with waterborne diseases. Diarrhea, often caused by cholera, is the world's second leading killer of children under five years old. And besides being a basic health issue, safe water is a women's issue. Women and girls are the water haulers of the underdeveloped world, so providing conveniently located sources of safe drinking water also helps keep girls in school.

Wealthier countries have been free of cholera for 150 years for one simple reason: we build water treatment plants and distribution systems. But the Western world is not helping the developing world to do the same. If nothing is done, Safe Water Network estimates that four billion people will lack safe water in the next ten to fifteen years.

As of 2019, Safe Water Network was serving more than one million people in parts of India and Ghana with water-purifying kiosks the size of a telephone booth. The key to their success has been charging customers a nominal but realistic fee—as little as five cents for 20 liters—to cover the cost of the kiosk, training its operator and staff, repairing and replacing parts, and educating consumers. If Safe Water Network is able to build a real network—if all the government agencies, charities, and nonprofits interested in safe water collaborated instead of competing—the model would bring safe water to those who lack it around the world. Shared billing systems could reduce providers' costs, freeing up more money for safe water. Shared strategic planning would allow safe water providers to close the gaps between systems and cover entire regions. Sharing scientific expertise could help solve one of the biggest problems: keeping safe water safe while it is distributed to users in jerry cans, on donkey carts, or in pipes. In some places, the rate of contamination can be as high as 60 percent—often because river water gets mixed in, containers are not properly cleaned, or children stick their hands into water jars in their homes. If water were tested for safety at every step of the way between kiosk and home, providers would know if and where water quality has been compromised, and where technology and consumer education could be improved.

Water safety is going to be the major issue of the next decade. With climate change, oceans are rising and safe drinking water is becoming increasingly scarce. Safe Water Network helps me use my experience with waterborne diseases to improve public health, and it's become one of my favorite activities.

My experiences with my own company, the Gulf of Mexico Research Initiative, and Safe Water Network have taught me that, deep down, I'm much more of a scientist than a capitalist. Many women scientists will enjoy the risk and potential profits of working in business. But I'm more interested in discovery, in figuring out ways to improve and save lives.

Over my six decades in science, finding ways to make the lives of others better and healthier has been a joy. Building networks and arming them with data to open the doors of universities, businesses, and government to women and underrepresented minorities has been another pleasure. My experience tells me that further change is possible. We can make the scientific community a better place for people to work.

chapter nine

It's Not Personal–It's the System

So far, this book has been a personal story, a look back over the years since I began my life in science in the 1950s. But here I must change direction and focus on the present context and conditions of women in science. When I talk with young women today, I hear them asking: *Has anything changed?*

To that, I would say: Yes. For example, both of my alma maters, Purdue University and the University of Washington, have had or have a woman president.

I also hear young women asking: *Is everything okay now?*

To this, I would say: No.

Deep down, many scientists are still convinced that the ability to do science is linked to the Y chromosome. We now have compelling evidence that the unequal treatment of women in science is an institutional and social problem. Women have the intelligence needed for successful careers. Innumerable studies document that the biological differences between men and women as they relate to science, mathematics, engineering, technology, and medicine are trivial or nonexistent. Girls have earned higher grades in school subjects—including math and science—for nearly a century, according to a massive historical analysis of more than a million boys' and girls' academic records in more than thirty countries (70 percent of the students surveyed were in the US). Other studies show that women don't underestimate their own abilities as much as men *over*estimate theirs. In addition, women today have the scientific degrees that should ensure success. In fact, women have earned more than half of the total number of science and engineering bachelor's degrees since 2000. And in the life sciences, they've earned about half the bachelor's degrees—and more than half the PhDs—for a generation, since the late 1990s.

Despite having both the scientific smarts *and* the scientific degrees, women are still not getting ahead. Once women earn their PhDs, they receive only 39 percent of postdoctoral fellowships, the stepping stone to a career in academic science, and have only 18 percent of the professorships. How can this be? It's not for lack of interest. The country has spent millions trying to interest girls and women in science—but as I've said before, women have always been interested in science. The fact is, women have been actively excluded from science for decades. And economically, we have all lost out.

The enormous cost of discrimination against women in STEMM affects us all. Nicole Smith, research professor and chief economist at the Georgetown University Center on Education and the Workforce in Washington, DC, explains why in her study.

Briefly, many women are interested enough in science, technology, engineering, or mathematics to obtain their undergraduate degrees in these subjects. (Smith's study did not include women with MD degrees.) Many of these women who go on to earn PhDs probably do so in other subjects (the US does not collect the statistics that would give us the exact number of women). However, a female PhD who does not use her STEMM background to work in a STEMM job pays a substantial penalty: she earns roughly $4,086 a year less than a female PhD with a STEMM job.

A large number of women are losing out. Approximately 140,000 women with STEMM PhDs take non-STEMM jobs and pay for it for the rest of their working lives. In comparison, only about 61,000 women with a STEMM bachelor's degree went on to earn a STEMM PhD and hold a STEMM job. If all those women were paid the same wages as men, their annual salaries would total $3.6 billion more than their current salaries.

Besides these women, the federal government also loses out to the tune of an estimated $772 million in tax revenue each year (assuming that STEMM careers would have put these women in a higher [25 percent] federal tax bracket).

And that's not all. The economy—and our national well-being—loses an estimated $4.6 billion per year because women working for lower wages invest less and have less to spend on restaurants, real estate, shopping, travel, and the like.

While $4.6 billion represents a mere 0.02 percent of the nation's GDP, it would be enough to end homelessness in Baltimore and Washington, DC—or provide pre-kindergarten for all American children.

The world economy counts on scientific researchers in the United States to start new technologies and enterprises. Since women outnumber men in undergraduate education and earn more than half of all science and engineering undergraduate degrees, it's going to be up to women to make the discoveries and advances we need.

So why are women continuing to be kept out of science?

Social scientists tell us that implicit bias (also known as unconscious bias) is to blame: our primitive brain and our primordial instincts make us assume with knee-jerk speed that whatever isn't like us is to be distrusted. As a protection against external threats, implicit bias makes sense. You're sitting in your cave roasting an animal over your fire when you hear rustling in the woods. You look up to see a creature with a totally different face—and your first reaction is "He's going to kill me!" Implicit bias helped the human species survive before civilization, so it's not wholly without purpose. Today, however, implicit bias is holding us back. It makes more sense nowadays to have open minds as we interact with those around us.

Implicit bias can be overcome with rational, careful deliberation, as Jay Van Bavel, a psychologist at New York University, has shown. Van Bavel organized an online game in which players saw someone stealing money. The thief was a member of either the players' in-group or another group. Players who swiftly—reflexively—decided the thief's punishment tended to punish malefactors from their own group more leniently than outsiders. But when the players took time to reflect before deciding, they were largely unbiased and punished members of their own group and outsiders equally.

There are tactics, then, that can limit implicit bias. The percentage of female musicians playing in the top classical music orchestras in the US went from 5 percent to nearly 40 percent in forty years after orchestras adopted the practice of auditioning new players behind a curtain so their sex could not be determined. Opening doors, though, does not guarantee equal positions or equal pay. Returning to the orchestra example, the principal slots—chief violinist, for example—remained 79 percent male as late as 2019. Nor did women players necessarily get salaries commensurate with those of their male peers. In 2019, a star female flutist sued the Boston Symphony Orchestra for equal pay; the suit was settled for an undisclosed amount.

Understanding implicit bias can also change hiring practices in science. In 2012, after a heavy round of layoffs at Yahoo!, Microsoft hired nearly every member of an all-male Yahoo! research lab. Microsoft Laboratory director Jennifer T. Chayes wondered how to change her new employees' culture. "Hmmm, guys," Chayes told them, "you seem to have lost a gender." Talking with each lab member individually, she explained how the impact of his particular research agenda could be increased by doing research in a more diverse lab. Chayes also had all the lab members attend a scientifically grounded workshop on unconscious bias. Thus prodded by their (female) boss, the men decided to look for candidates across a wider landscape and changed some of their interviewing practices. Before a candidate appeared, each lab member chose one of the interviewee's papers to read so that the group formed an impression of the candidate's work before seeing him or her. Instead of sharing their written opinions about a job candidate immediately, the lab eliminated groupthink by waiting for everyone's evaluations to be written up before discussing their opinions. Within two years, the lab went from zero women to 30 percent women—and it was the male researchers who made the change.

But not all men are receptive to challenging biases. When microbiologist Jo Handelsman tried to talk with male scientists about unconscious bias against women in science, many of them couldn't believe they might be part of the problem. "We don't do that," they told her. "Science is objective. We only hire the best, and we know it when we see it."

"I've probably heard that a thousand times," Handelsman told me. And yet, as a postdoc pointed out to her, scientists run randomized double-blind experiments because they know they can't be unbiased about their data. How could they expect to be unbiased about everything else?

Handelsman and her team at Yale decided to perform an experiment on scientists themselves. She convinced 127 scientists in biology, chemistry, and physics departments at six leading research universities across the country to evaluate a job application, without telling them the purpose of her questionnaire. Ostensibly, the application came from a recent graduate seeking a position as a laboratory manager, and all the applications were identical—almost: half the applications were signed "John," and half were signed "Jennifer."

The disturbing results revealed the depth of unintentional discrimination in science. Both men *and* women scientists judged the male applicant to be more competent than the female applicant with identical qualifications. More faculty members said they'd hire John than Jennifer. They'd offer Jennifer less support. And they'd pay her almost $4,000 less per year ($30,238.10 for John and $26,507.94 for Jennifer). All across the board—no matter their age, sex, scientific field, or tenure status—faculty members preferred John. The results were "very threatening to people who pride themselves on their objectivity," Handelsman said.

She herself was stunned by the results—especially by the fact that scientists would offer a woman less help. As Handelsman said, "Every time the woman goes in for advising, to ask questions, to learn about a program, find a summer research project, for the two-minute question after class as to who gets to go on a field trip—every time, she gets less support. Multiply it out, and you realize the magnitude of the potential impact on the confidence and reinforcement that women are getting."

Each discriminatory episode might seem insignificant, but repeated over time like compound interest, even molehills become mountains, Hunter College psychologist Virginia Valian showed in her 1998 book *Why So Slow?* A woman can begin her career earning the same salary as a man, but ten or twelve years later, the man will be almost a full academic rank ahead. As Nobel Prize winner Elizabeth Blackburn said, "A ton of feathers still weighs a ton."

Since Jo Handelsman's John and Jennifer study, a deep-seated bias against women in science has been documented at almost every level, from Nobel Prize winners down to undergraduates. In 2014, MIT biology graduate student Jason Sheltzer and Twitter software engineer Joan Smith showed that the more decorated a male biology scientist was, the fewer women he trained. And if the professor had won a Nobel Prize, membership in the National Academy of Sciences, or a grant from the Howard Hughes Medical Institute, their postdocs were 90 percent more likely to be male than female. The worst of it is that universities tend to hire their junior faculty members from these elite male feeder labs.

A year later, in 2015, Nobel Prize–winning biochemist Sir Richard T. "Tim" Hunt told a conference of female science journalists why this is. "Three things happen when they [women] are in the lab," he said. "You fall in love with them, they fall in love with you, and, when you criticize them, they cry." Hunt suggested that to avoid this problem, men and women should work in gender-segregated labs. He then made matters worse by telling BBC Radio 4, "I have fallen in love with people in the lab and people in the lab have fallen in love with me, and it's very disruptive to the science because it's terribly important that in a lab, people are on a level playing field." These comments are sheer nonsense, but they illustrate the type of bias many women have had to deal with. Hunt later apologized for his remarks and has since proved to be an active supporter of women's issues.

The news that scientists are as biased as ordinary mortals reached the national media several years before women's stories of sexual assault by filmmaker Harvey Weinstein began appearing in the press in October 2017, taking the #MeToo movement mainstream. Science scandals featured male celebrity professors at Yale Medical School, Berkeley, the American Museum of Natural History, the University of Chicago, Caltech, the University of Washington, and Dartmouth. The professors involved worked in medicine, astronomy, anthropology, molecular biology, astrophysics, and brain science. Most were winners of big grants. And in most cases, administrators had ignored prior complaints about their behavior, sometimes for years. After media publicity and sometimes even faculty protests, these men lost grant funding and, in some cases, their jobs.

Unfortunately, waiting for a new generation of presumably more enlightened young men is not going to rid science of bias. When 1,700 male undergraduate biology students at the University of Washington were asked to name the strongest students in their class, the men consistently underestimated their female peers by three-quarters of a grade point, even though the female students actually performed better in the class. "It's like believing a male with a B and a female with an A have the same ability," said Sarah Eddy, who co-led the study. In addition, gender bias among the male students was an estimated nineteen times stronger than bias among females—so much that Eddy doubted she and her colleagues could do much to mitigate it. "As science instructors at the college level, you can only affect so much," she said. "There's been at least eighteen years of socialization before that." It's the system, not the women, that needs to change.

Bias against women in science certainly isn't limited to young men at the University of Washington. Studies show that men publish more papers, but women's papers are cited more and so are more influential. Men think computer code written by women is better—as long as they don't know the coder was female. Letters of recommendation for women are shorter and stress that the applicants "work hard" and are "diligent"; letters about men use words like "outstanding" and "superstar." In Sweden—supposedly a world leader in gender equality—in order for a woman to be considered equally qualified as a man for a scientific job, she must publish three more papers in one of the world's most prestigious science journals (like *Nature*) or twenty more papers in the top journals in her field. Men talk shop with their male STEMM colleagues but make small talk with female coworkers. And when the men discuss research with other men, they rate women as less competent. In some fields, like economics, women get zero recognition for papers they coauthor with men; Heather Sarsons, a Harvard economics grad student, began a paper about her discovery of this phenomenon with the words: "This paper is intentionally solo-authored." Two women who submitted an article to the journal *PLOS ONE* in 2015 were told to improve it by getting "one or two male biologists" to sign on as coauthors. Incidentally, the reviewer rejected the paper because (and I quote) "the qulaity [*sic*] of the manuscript is por [*sic*] issues on methodologies and presentation of resulst [*sic*]." To its credit, *PLOS ONE* later removed the reviewer and an editor who had been involved in the decision.

The list goes on and on. Female and other minority executives who promote diversity are penalized with worse performance ratings; males who do the same are not. Male faculty in STEMM are more reluctant than women to believe research about gender bias. And if instructors in training sessions about bias don't explicitly state that stereotyped biases are unacceptable, the training can actually increase the problem. Years after MIT women rose up to protest being marginalized and offered fewer resources, three women scientists denounced similar problems in suits against the elite Salk Institute for Biological Studies. The 2017 lawsuits blasted a male-dominated culture of marginalization and hostility, lower wages and research funding, and less lab space. A leading male scientist at the institute was placed on administrative leave amid allegations of sexual assault made by eight women. He resigned two months later; the suits were settled the following year.

Bias translates into dollars. On average, the NIH gives bigger grants to men, $41,000 more per grant than what women receive. The gap between NIH grants for males and females is even larger at top universities like Yale and Brown: $68,800 and $76,500, respectively. It also takes female postdocs a year longer than men with equivalent publications to get a grant from the NIH.

Change is not going to be easy. One significant action that scientific leaders can take to combat bias in their fields is to issue speaking invitations to women who deserve them. Speaking invitations are as vital for a scientist's career advancement as publishing their work; both show promotion and tenure committees that the scientist's work is respected by the research community at large. Remember how Mary Clutter at the NSF a quarter of a century ago tried to reject funding for scientific meetings that hosted only male speakers? Or how in the 1990s, it took Barbara Iglewski the entire decade to get women onto the committees that accept articles for publication in the American Society for Microbiology's journals?

In 2014, two brave leaders tried to get more women invited to speak at important meetings of the ASM. By then, the society had more than

39,000 members, roughly half of them women. But Jo Handelsman and Arturo Casadevall gathered data showing that all-male committees organized half the sessions at the ASM's big meetings in 2010, 2011, and 2013. A third of those all-male committees invited only male speakers, even though mathematician Greg Martin calculated the odds of a panel being "randomly" all male as astronomically slim. "They don't just happen," Martin told *The Atlantic*. Handelsman and Casadevall's statistics also estimated that including just one woman on each of those all-male inviting committees would increase the number of women speakers by roughly 72 percent.

The first year that Casadevall sent these statistics to each committee, nothing happened; the number of all-male sessions did not change. Then he tried a personal approach, pleading face-to-face with committee members to "do better" and avoid all-male sessions "except under extraordinary circumstances." Apparently that worked. A year later, in 2015, roughly one hundred more women gave talks at ASM meetings than in previous years. In all, 48.5 percent of speakers were women, basically matching the percentage of women in the profession. For the first time in ASM history, the general meeting had achieved gender equity among speakers. Casadevall noted, with some satisfaction, that the "near eradication" of all-male sessions showed that change can occur in a relatively short time. Four years later, NIH director Francis Collins announced that he would speak only at conferences "where scientists of all backgrounds are evaluated fairly for speaking opportunities."

Transformation is possible. Seventy years ago the University of Maryland refused to admit African American students. Now I'm proud to say that the university ranks eighth in the United States for graduating black PhDs. More black students graduate from the University of Maryland with PhDs in mathematics and statistics than from anywhere else in the country.

The United States as a whole has undergone massive changes in attitude within much shorter periods of time. Views on smoking, drunk driving, the Vietnam War, and LGBTQ rights have shifted dramatically, sometimes within a single decade. It took enormous effort on the part of a great many people, but change did happen—and it happened relatively rapidly. Today, attitudes about women, African Americans,

Latinx, and LGBTQ people in science are beginning to change, too. But there's still a distance to go.

Consider the National Academy of Science, an honorific organization formed in the mid-nineteenth century to advise the country on scientific matters. Because current members decide which new members to nominate—and because science has been predominantly white and male for more than 150 years—the academy is now 83 percent male, with an average age of seventy-two. Yet this, too, is changing. About thirty years ago, the academy formed a Committee on Women in Science, Engineering, and Medicine (CWSEM), primarily to elect more women into the academy. The committee was given no resources and was scheduled to gather during the academy's annual meetings at seven a.m. on Sunday in the former basement cafeteria of the academy. The basement had exposed overhead pipes and poor lighting and no accommodation for slides or microphones. We were left to buy our own breakfast in the academy cafeteria. About ten years ago, some of us complained about this blatant sexism, so the governing council moved us upstairs—although the meeting remained scheduled for seven a.m. on Sunday. We were allowed only one hour because the upstairs room was needed for "more important" meetings.

When I was elected to the academy's governing council, I publicly reminisced about the "honor" of being moved upstairs from the basement, though still in the early morning twilight; the next year, we were scheduled in a meeting room for a noon lunch. Then, in 2014, CWSEM invited a speaker: Jo Handelsman, by then a federal appointee to the White House Office of Science and Technology Policy. Our luncheon was moved to the Great Hall—though we still had to pay for our meals. The following year, we invited Secretary of Homeland Security Janet Napolitano to speak, only to find the Great Hall's tables filled mostly with men, who were also called on to ask the first questions. No tables had been reserved for members of CWSEM. Maxine Singer, one of the founders of the committee, was seated at the back of the room next to the exit, while Kitty Didion, the National Academy of Sciences' staffer on the committee, couldn't even get a ticket. She had to sit outside, listening as best she could through the closed door. After that experience, the question we have to ask is, *What is wrong with our country, our society, and our supposed leaders? How much longer are we going to have to do studies and talk about our problems?*

A rigorous report documenting the facts was needed. All three National Academies—Science, Engineering, and Medicine—would have to work on it together. And it would have to be prepared by a committee of very accomplished, highly respected, credible people who could not be dismissed as whiners. Academy president Marcia McNutt and CWSEM members (I was committee chair at the time) assembled a remarkable group for the study: co-chairs Sheila E. Widnall, who dealt with the Tailhook scandal as secretary of the air force, and Paula A. Johnson, president of Wellesley College; former congresswoman and ambassador Constance Morella; and psychiatrists, lawyers, and businesspeople. The academy's first-rate staff spent more than two years commissioning studies, gathering data, and tabulating and interpreting what they found.

The report had clear findings:

- We can't depend on the law to address or prevent sexual harassment against women in STEMM careers, because that approach hasn't solved the problem. The law only requires institutions to demonstrate that they have a sexual harassment policy in place, and that they're following it—not that such efforts are proving effective at reducing or preventing harassment. Our academic institutions need to treat the law as the bare minimum required; only system-wide changes to culture can solve the problem.
- Sexual harassment is a form of discrimination that consists of three types of harm:
    - Gender harassment—the use of denigrating language, jokes, comments, or nonverbal behavior (like the circulation of degrading images)—that communicates that women do not belong or do not merit respect. It is the most common form of harassment, but many people do not realize that gender

    harassment is a kind of sexual harassment. When it persists, it can be as damaging to a woman's success as sexual coercion, and where it is tolerated, other types of sexual harassment are more likely to occur.
    - Unwanted sexual attention, such as unwelcome sexual advances and sexual assault.
    - Sexual coercion, when favorable treatment is conditional on sexual compliance. This is also known as quid pro quo harassment.

- The biggest predictors of sexual harassment in an organization are that the group has traditionally been dominated by males, who still hold the power; that sexually harassing behavior is perceived to be tolerated; and that faculty and trainees are often isolated for considerable periods of time in labs, field sites, clinics, and hospitals.
- An extensive analysis of two university systems showed that between 40 and 50 percent of medical students have experienced sexual harassment from faculty or staff. More than a quarter of female engineering students and about 20 percent of female undergraduate and graduate students in science have, too.

Fortunately, the committee's report was ready in time to be published during the rise of the #MeToo movement and brilliant spotlight was cast on sexual harassment and its devastating effect on women's careers. Its findings were covered by more than 100 news outlets, at home and abroad, including the *New York Times*, the *Washington Post*, NBC News, and PBS. Of the hundreds of reports the academy has published, "Sexual Harassment of Women" is in the top 1 percent of most requested press products to date and has proven to be a milestone in the fight for women's rights in society.

Months before the report was published, the National Science Foundation put its revised terms and conditions in the Federal Register for comment. The NSF, the lead funder of the study, now requires institutions to report accusations of harassment under investigation to the foundation. The National Academy of Science, National Academy of Engineering, National Academy of Medicine, and the American Association for the Advancement of Science can now revoke the membership of elected members in cases of proven scientific misconduct or serious ethical breaches, such as sexual and gender-based harassment. The American Geophysical Union recently defined harassment as a form of scientific misconduct. And in February 2019, NIH director Francis Collins apologized for the agency's past failure to "acknowledge and address the climate and culture that has caused such harm" and the agency replaced fourteen grant recipients who had engaged in harassment.

Still, the battle against discrimination is far from won. How can we achieve true equality within the scientific enterprise so that men and women can thrive and compete as equals? I have some ideas about how to do just that.

chapter ten

We Can Do It

I conclude this book by passing on to women and their allies some ideas for steps they can take to help reform American science. These suggestions are based not just on my six decades of experience in science, but also on the experiences of other women scientists and recent scholarly research. My advice is aimed particularly at young women contemplating a career in science, but it's also for parents, teachers, institutions, and lawmakers concerned with the serious scientific challenges of our time.

My suggestions are predicated on several beliefs:

- All men and women deserve equal treatment in school, in the laboratory, on the job, in career advancement, and in their personal lives.
- We do not need to cater to women in science. We need only give women an equal chance to achieve.
- The best of 100 percent of the population will always be better than the best of 50 percent of the population. Once all the talent in our country can compete on a level playing field, decisions about who to hire and who to support can be made on the basis of brains and ability, not gender, ethnicity, or national origin.
- Attitudes will have to change. Relying on the law to

> achieve equal opportunity has not worked. True equality for women in science will require broad societal change.

The twenty-first century has already brought about great challenges, and more are sure to come. Climate change and rising sea levels will exacerbate food and safe water problems for the ten billion people expected to populate our planet by 2050. Global security, social stability, and economic prosperity will demand every talent and the ability of every citizen in the world, regardless of gender, ethnicity, or national origin.

Here's what we need to do to prepare:

FIRST, THINK POSITIVELY

Just the other day, I was sitting in a meeting, clearly able to see our goal. But everyone around me was thinking only of reasons not to do what needed to be done. Every comment was why it wouldn't work, why it wasn't the right time, why it didn't make sense . . . But now isn't the time to look down. Now is the time to look to the horizon and over the next mountain.

Persevere. A scientist who perseveres thrives—for two reasons. First, because the establishment will try to resist change. And second, because nature does not reveal her secrets easily. The same skills that will help you navigate the system are those that will support you when experiments don't yield the results you'd hoped or planned for, and you have to figure out your next steps.

Know your goals, because the way to achieve them may not be a straight line. Your path may be longer and harder than you expected and require circumnavigating obstacles. But if that path gets you to the other side, take it. The best way to change a system is to be as successful as your talent allows, because then you can change the system from within.

• •

To quote Florence Haseltine, who helped convince the NIH to form an obstetrics-gynecological program: "Don't take 'no' for an answer. 'No' just means the other person isn't going to help you. It doesn't mean you can't do it."

If you think this advice is dated, listen to Crystal N. Johnson, an African American environmental microbiologist and tenured associate professor at Louisiana State University. Johnson came from a desperately poor family in rural Mississippi. She ended her first undergraduate semester at Tulane University with a 0.5 GPA and was put on academic probation. "It shattered me," Johnson says, "but it lit a fire in my belly." She studied intensely, but she also collected a group of online mentors whom she could email for advice and support (she calls this "e-mentoring"). She pulled up her GPA, graduated with high honors, and was admitted to graduate school. Johnson counsels students to be flexible. If something doesn't work, try another way. Pan negative comments for gold. And if you're introverted, as she was, she says, try to overcome your fears.

• •

"You'll be surprised how far some humor can carry you."

—Alice Huang

TELL GIRLS THEY CAN DO SCIENCE AND TECHNOLOGY

"Teach girls to stand up for themselves. If the world were completely fair, we wouldn't have to do that but, as of right now, it is not fair—so we do."

—Shirley M. Tilghman, former president of Princeton University

• •

Teach boys as well as girls to do their own laundry, cook, and clean up after themselves. Don't send kids out into the world thinking the world owes them anything.

• •

As I mentioned earlier, one of my favorite National Science Foundation programs paid graduate students in STEMM to teach science five hours a week in a middle or high school classroom. Help support a program that will bring scientists and engineers—especially young women scientists and engineers—into a school. It's hard to picture yourself as a scientist or engineer if you've never met one.

••

Parents and teachers: Encourage girls to believe they are smart, even preschool girls. By the time girls are six years old, they have already been conditioned to believe they aren't as smart as boys. And six-year-old girls have already begun to avoid activities if they think they're for "really, really smart" children either because they are conditioned to think that isn't a desirable quality or because they don't think they'd qualify. But studies show a little bit of encouragement can overcome those fallacies.

Start encouraging girls' confidence in their ability to learn math and technology even *before* they're in preschool—and even if you know nothing about the subjects yourself. A Google study of 1,600 men and women in the United States concluded that encouragement from family members and teachers—whether they know anything about computers or not—is the most common reason young women study computer science. Other reasons they succeed include an interest in puzzles, problem-solving, and exploration, as well as childhood participation in computer science courses and activities.

In 2012, women made up only 26 percent of computer science and mathematical science professionals in the United States—though the country's need for computer science professionals severely exceeds the labor supply.

••

The website of the National Center for Women & Information Technology (NCWIT) lists ten ways families can encourage girls' interest in computing. NCWIT is supported by Apple, Microsoft, Bank of America, Google, Intel, Merck, AT&T, the Cognizant U.S. Foundation, and the NSF.

••

High schools should be teaching computer science, not just computer literacy (using computers for word processing, creating spreadsheets, online shopping, and such), which is what most schools teach. Computer science involves writing programs that carry out a task. It combines logical reasoning, problem solving, and adapting new technologies to design solutions to problems in science, math, the social sciences, and the arts. Pick a problem that interests you: saving a river, curing a patient, building a new school . . . Then start learning a computer language like Python or JavaScript and design a solution.

• •

If you have a choice, pick a high school that focuses on academics. Find a school where everyone (*everyone*) is expected to succeed and no one gets away with less than their best work. The world is complicated, and children, especially girls, must be educated to the maximum extent of their abilities.

Ask school authorities whether their graduates—girls *and* boys—become scientists, engineers, doctors, lawyers, businesspeople, entrepreneurs, and socially conscious public servants.

Try to find a school that teaches discovery science. This is science taught by doing, not memorizing. Discovery science stimulates creative thinking, questioning, learning, and especially understanding.

TRAIN YOURSELF

Form or join study groups that include other girls or women. Study groups enhance learning: what you don't understand, someone else will. Talking about a problem can help you find a solution, and friends can help when you feel close to "burnout" or disappointed by some development adverse to your career plans.

A good education includes studying other languages and cultures, literature (the classics as well as contemporary fiction, poetry as well as prose), and mathematics. Yes, mathematics, mathematics, and—in case you didn't hear it—mathematics. The nonsense about girls not being good at math is totally outdated, wrong, and ridiculous. Study non-STEMM disciplines, too. "The world would be a better place if we could open our young minds to the rigors of disciplines beyond STEMM," wrote Satyan Linus Devadoss, a fellow of the American Mathematical Society and a professor of applied mathematics and computer science at the University of San Diego. I agree with him.

The humanities deal with "gender and race, beauty and acceptance, truth and power," Devadoss wrote. "These problems are . . . far more difficult than rocket science." The humanities teach us ways to study complex frameworks by the "close readings of texts, unpacking the subtlety of rhetoric, embracing multiple ways of knowing, and valuing experiences acquired through arduous history."

Learn how to write and speak effectively. In the short term, colleges love students who can talk enthusiastically and knowledgeably about the novels and poems they read, as well as the science they're interested in studying. In the long term, reading and writing well will help you think better and will enrich your personal and professional life. Clear and effective communication is essential for success in any field.

As a student, I took courses in poetry, creative writing, and literature, and I can assure you that they helped me see problems and issues from different perspectives. The contributions of social and behavioral scientists—and especially writers and artists—are needed more than ever before to help us address the challenges of today's complex world. Individuals with unidimensional personalities may think they're successful, but they're actually socially handicapped.

• •

Stay in school. Women need two extra years of postgraduate education just to earn as much as men, reported William R. Doyle of Vanderbilt University in a 2008 study. More women enroll in higher education than men; that's true at both undergraduate and graduate levels. But Doyle's study found that young adult males had higher median earnings than young adult females with the same level of education.

• •

Explore science and technology widely. Half—or more—of what you learn as a student will be outdated by the time you graduate. A field that's popular today may be obsolete by the time you enter the job market, while a field that will be exciting in years to come may not even exist yet. If you explore enough to discover your own genuine interests, you're more likely to persevere against the odds and be successful.

Take mathematics and computer science classes—starting as early as possible but at least in high school, if not before. Read about the subjects on your own or take extra courses, so you'll be as prepared as possible in college. No matter the scientific discipline, the old scientific methods are incorporating new techniques made possible by high-speed computers and probability mathematics. Modern biology is heavily computerized, including modeling, simulations, and data mining. And in big interdisciplinary science, computer programming and statistics are valuable tools. Complex, global systems are composed of enormous amounts of big data and follow many different pathways. Figuring out how they intersect will require machine learning (artificial intelligence) and powerful visualization tools.

• •

Rating universities and colleges by how well they treat women—and putting the ratings online—would help level the playing field for women. Ideally, there would be an online spreadsheet listing all universities with very specific categories about their treatment of women, especially for postgraduate study. Which universities provide or subsidize daycare and after-school care? What's the gender balance of the faculty and administration? Can the tenure clock be stopped? Which departments have women faculty, how many, and at what rank? And do women's salaries compare with those of their male peers? Harassment problems and policies should be included. All these parameters are best put on the web so everyone—students and faculty, parents and taxpayers—can know a school's characteristics and use that information to decide if it's the right place or one to avoid.

• •

Before college, get some experience doing research or working in a lab. Otherwise, you won't have a real feeling for what it takes to be a laboratory scientist.

"I'll tell you what I see a lot around here," says distinguished professor Tamar Barkay at Rutgers University. "Typically, every spring, I get numerous requests from high school students who want to come and spend the summer with us. Many of these students are what I call science junkies. They take every opportunity, no matter what kind it is, to spend their free time during the school year or summertime to expose themselves to more science. They take courses, spend time in labs, do special projects, and take every opportunity to participate in science competitions at different levels."

Barkay advises going online to see what resources are available. Perhaps a nearby college allows high schoolers to enroll in classes, or a guidance counselor or reference librarian can provide information about a summer internship, a local business with scientific aspects, or other opportunities to immerse yourself in science before college.

• •

Follow *Discover* magazine, *Scientific American*, *New Scientist* magazine, and/or the *New York Times*'s Tuesday science section. All these are available online, and if a publication is behind a paywall, schools often have subscriptions that students can access.

• •

Go on field trips, but gather information before you go. Research trips are required for classes, majors, and careers in many social, life, and earth sciences. Because many male professors claimed living conditions in the field were far too primitive for delicate females, women fought for decades to join these trips. Now that women outnumber men in field-based sciences at undergraduate and graduate levels, female trainees routinely go on research expeditions. Still, check with your school or sponsoring organization about its code of conduct and sexual harassment policies for field trips before joining a field trip or scientific excursion.

GO TO GRADUATE SCHOOL

Before choosing a graduate school, get advice from several mentors. Do not rely on one person to advise you. Professors you admire may not have the latest information on grad school programs. And don't pick a place just because you have a potential mentor there. Mentors can become seriously ill, change jobs, or even die.

• •

"I would recommend to young women that they do not accept admission to graduate departments which do not have at least a couple of women faculty members, preferably tenured, and several women graduate students . . . A woman is less likely to make it [in science] if these supports do not exist."

—Physicist Fay Ajzenberg-Selove

• •

Consider a PhD program that includes computer science and statistics training. Financial and management training would also be useful.

• •

Once you're in graduate school, choose your thesis advisor carefully, because this is one of the most important decisions a scientist can make. Try to find a good scientist with a record of helping women go on to have successful careers. Here are steps to take before choosing a PhD advisor, from developmental biologist Constance L. Cepko of Harvard:

- Look at the quality of the science being done in the lab.
- Pick someone who will train you to be a good scientist and not just use you as lowly paid labor.
- During your first year in graduate school, rotate through three labs, spending a few months in each. Get a sense of how students are supported. Are they encouraged? Helped? Talk to many people in the lab, not just the teacher's pet. And don't select an advisor just because

you liked the postdoc you worked with during your rotation.

- Sit down in the faculty member's office for an hour. Do you feel comfortable? Are you having a two-way, respectful conversation, not just about data but about your future plans and career? Were your opinions heard? When you leave the office, ask yourself, "Would I want to go there again?"

••

"Science is different from other professions because there is no statute of limitations on a mentor's influence over his advisee's future," explains geophysicist Marcia McNutt, president of the National Academy of Sciences. Your PhD mentor will be your reference for years to come when you're under consideration for grants, conference talks, and publications. A good mentor will also protect graduate students from others who may want to purloin their ideas.

MAKE A CAREER

My greatest good fortune was having a life partner who was understanding and supportive in every possible way. My husband, Jack, and I went through graduate school together, raised our family as a team, and always made sure we spent time, however limited, with our children. This was not easy, but it was a priority for both of us. It definitely made our life together loving and joyful.

Know that it is, indeed, possible to be a scientist and a mother. Both of my daughters have successful careers. Stacie is an MD/PhD specializing in developmental pediatrics and palliative care, all while raising three children with her supportive husband. My other daughter, Alison, is a talented botanist who delights in her research on plant evolution and is an authority on California's wildflowers and parasitic plants.

••

Academic science is more family friendly than many other professions, Cepko says. "We work hard but we have incredible freedom. We make our own schedules. If we can get grants, we can do something we're really interested in. If you really love science, there's nothing like it. If you're a smart young woman and work hard, there are lots of opportunities."

• •

Don't agonize about getting published in *Science*, *Cell*, or *Nature.* They're for "the connected," said biologist Randy Schekman as he collected his Nobel Prize in 2013 and announced that his lab would no longer send research papers to them. Pressure to publish in top-tier journals—whose editors want "eye-catching" articles—encourages researchers to pursue trendy fields of science instead of genuinely important problems. Schekman supported *eLife*, a peer-reviewed, open-access journal for the biomedical sciences founded in 2012 by the Howard Hughes Medical Institute, the Max Planck Society, and the Wellcome Trust. There are many good, peer-reviewed journals being published.

• •

Don't discount work done by little-known scientists just because they're not working at a major research institute. Judge information on its merits, not its pedigree.

• •

If you're isolated—the only woman in a department, the only member of your specialty in your college, the only single mother, whatever—go online and find kindred souls. Then set up a regular schedule to meet virtually twice a month and face-to-face once a year. Don't be alone.

• •

Men are four times more likely to ask for a raise than women—and when we do ask, we ask for 30 percent less, according to Linda Babcock, an economist at Carnegie Mellon University and the coauthor of *Women Don't Ask*. Be realistic. Make an honest estimate of the position involved and its responsibilities. Confer with an advisor or mentor to get an estimate of appropriate salary ranges. Then start with the high end in your negotiations with your employer.

••

No one gets a prize without first getting a nomination. Ask people to nominate you for prizes that you or your mentor believe your work merits—men do this all the time. Offer to help by providing the information needed for the nomination and a draft of a possible nomination letter. If you're asked to nominate someone, ask if they can supply draft nomination forms to help you get the facts right. (Some men pair off and alternate writing nominations for each other—a practice of which I do not approve.)

••

Make a habit of empowering other women. A token individual—whether a woman in a male group, a black among whites, or a man in a female kaffeeklatsch—is highly visible within a group and likely to become stereotyped, which makes assimilation hard. If another woman joins the group, Harvard Business School professor Rosabeth Moss Kanter explains, you'll be lumped together, whether you like it or not. So don't undermine one another—support each other. This can be as simple as a friendly conversation or telling her privately that you appreciated something she said or did.

HOW MEN CAN HELP

To those men who are already good allies, proselytize to other men. Speak up about microaggressions every time they occur until other men are aware of what they're doing and think twice about what they say and how they say it. Courtesy and kindness are not signs of weakness, and misogyny is a strong indication of a lack of character. Know that, as an ally, you are deeply appreciated by all of us.

••

Alice Huang says, "I'll never raise my hand again to be the secretary or the organizer of celebrations for a group . . . Let the men do it for a while."

••

Don't comment on anyone's appearance or clothing—those things are irrelevant. They speak to class, status, and money, not to competence or talent.

••

"I have a lovely set of colleagues, but in meetings, they become aggressive. If they don't agree with something I've said, they shout at me until I give in. They don't have the social skills to understand that that's unacceptable. I've never had a female colleague scream at me. Yet nobody senior says it's unacceptable."

—Anonymous

HANDLING TOXIC ENVIRONMENTS

At some point in your life, you may find yourself in a toxic environment. In the best of all possible worlds, the twin goals for any woman in such an environment would be to stop the harassment *and* protect her education or career. That may not be possible. Realizing this, the question becomes: Should she make a report, and if so, how and to whom? The answer depends on the woman and the situation.

Even if you don't want to report an incident of harassment, document it immediately. Be specific. Don't say, "So-and-so harassed me." Say exactly what he (or she) said or did, as well as where and when it was said or done. Then either have the document notarized or email it to trusted friends the same day. Sexual predators often repeat the same behavior until they're caught. The fact that you documented your experience could help another woman who wants to report her own harassment incident later. Evidence of pervasive and repeated behavior is required to prove the existence of a hostile work environment.

Go to university authorities. Go to your department chair accompanied by trusted peers who support you, even if the problem is yours alone. If you're unsure whether the department chair will help, ask around: Is he or she regarded as reliable or part of the problem? If talking to your department chair doesn't work—or seems too risky—proceed upward to the ombudsman, dean, provost, president, and regent. From an institutional perspective, harassment is a disaster, if only because of the financial liabilities; the administration should pay attention. That being said, hiring a lawyer may become necessary if the situation is persistent or egregious.

• •

A distinguished woman scientist in the middle of her career made a confidential, after-hours appointment with me to talk about problems she was having with a world-famous (male) scientist at work. She asked me what she should do.

I know the man. Like too many other senior males, he does not believe women can be intellectually equal to men—and *he does not think bullied women will fight back*. And for the most part he's right.

"No one should put up with appalling behavior, and you must leave," I told her. "But leave *with your career intact*. Make a list of what you need to accomplish to get a good job elsewhere, give yourself up to a year to accomplish those things, and then, at the end of that year, go." Which she did, landing in a much better place. Unfortunately, the man's bad behavior was never dealt with; it should have been.

A young woman at Tennessee Tech University brought a harassment complaint to her department chair, who quietly asked a colleague, Professor Sharon Berk, a former PhD student of mine, if the young woman could finish her thesis in Berk's lab. Which she did.

While I wish it were possible for every perpetrator to be disciplined immediately and without repercussions for the victim of their abusive behavior, such predicaments can have nuances.

The National Academies' report on sexual harassment calls for escalating disciplinary consequences depending on the severity of the abuse. Not all behavior should be grounds for dismissal, but abusers should certainly be held accountable, before the situation worsens.

TAKE THE OPPORTUNITIES OFFERED BY TECH

History shows that as fields go from being male dominated to female dominated, the wages decrease. It's happening in computer science, and it will most likely happen next in machine learning and robotics. Data science, artificial intelligence, and new visualization tools are the future, and women interested in computer science should be learning to master them now.

• •

Excluding women from science and technology can be highly detrimental to the country. During World War II, British intelligence engineered the first large-scale programmable electronic computer, known as the Colossus. At the UK's Bletchley Park, codebreakers, almost exclusively women, used the Colossus in conjunction with rudimentary computing machines and handwritten techniques to break German military codes. As computers came into wider use, the female scientists operating the machines were required to train male replacements who knew next to nothing about math or computers. The women, who knew a great deal about both, were fired when they were done passing on their expertise, and the British computer industry fell way behind that of the US.

• •

Patenting is often the first step to a consulting job or appointment to a remunerative post on a corporate scientific advisory board, and male academic life scientists patent twice as many of their discoveries as their female colleagues. Most universities today have offices to help scientists patent their work—women should visit these offices and make use of them.

• •

Serving on a corporate or nonprofit board of directors can open up scientific opportunities and sources of funding in business and industry. Women need to build résumés specifically aimed at gaining membership on corporate boards. You might begin as a volunteer board member for a local organization, then progress to regional boards, and finally send your résumé to recruiters who locate board members for the corporate world.

BECOME THE PUBLIC FACE OF SCIENCE

Science and technology are pillars of the modern world, but congressional support for science is waning. More than 50 percent of the US population does not believe in evolution, and the wave of parents choosing not to vaccinate their children ignores one of the biggest medical breakthroughs of the past century. It is the responsibility of every scientist to explain her work to the public and give it a human face. Unless the American people can be convinced that basic scientific research should be a higher priority, the country itself is going to be in serious trouble. Every woman scientist must become a spokeswoman for science. Learn to explain your field effectively to the general public and the legislators who represent you. Hammer home the message that more women in science and engineering means better science and engineering—and a better world. Learn to lobby. Consider running for public office.

The most important issue a woman scientist can speak up about is childcare: universal, affordable, high-quality childcare. Nearly half of the female scientists in the US leave full-time science after their first child is born. In comparison, 80 percent of male postdocs and female postdocs without children stay in science. Women who do become faculty members tend to have fewer children than their male colleagues—or have fewer than they want to have.

A bipartisan bill to finance universal childcare passed Congress in 1972, but President Richard Nixon vetoed it. His veto cost us two generations of women scientists and their discoveries—and two generations of children who could have had a safe place to stay during the day while their mothers worked and the opportunity to interact with other children and learn early communication and logic skills. Until Congress gets its act together and finances childcare for everyone, employers will have to do so.

Many universities now offer childcare facilities for faculty, postdocs, and graduate students and subsidize childcare for undergraduates and staff with children. They also often try to help their faculty members by offering their college-aged children reduced tuition. Sharon Berk of

Tennessee Tech University asks, "Why not let young faculty members use the money instead for day care?" should they prefer to.

• •

Protecting our state colleges and universities is a women's issue. Women need affordable higher education near home. Despite the strides toward career and household equality women have made in recent decades, by and large we still have more modest incomes and heavier family obligations than men.

• •

Advocate for more money for science research. When money for research dries up—as it has recently—young women are the first to lose their funding. Grants tend to go to low-risk projects conducted by established scientists and engineers.

At the NIH, the main funder of medical research in the US, the proportion of all grant funds awarded to scientists under the age of thirty-six fell from 5.6 percent in 1980 to 1.5 percent in 2017. "How successful would Silicon Valley be if nearly 99% of all investments were awarded to scientists and engineers age 36 years or older, along with a strong bias toward funding only safe non-risky projects?" Bruce Alberts, the former president of the National Academy of Science, and Venkatesh Narayanamurti editorialized in *Science*.

"From an economic sense, it's really irrational," supercomputing scientist John West of the Texas Advanced Computing Center at the University of Texas in Austin has said. "If you've been funding these people all along and they still haven't given you the new internet, guess what? It's unlikely they're going to. So why don't we try funding some new people, and then maybe you'll get a surprise."

Tell your legislators that underfunding basic research harms American education and our global competitiveness. Amy Millman, founder of Springboard Enterprises, occasionally teaches a graduate business course in Washington, DC. One day she looked around and realized she was the only American citizen in her classroom. All the students were foreign-born. Congressional and state legislators aren't supporting public higher education, so universities are raising tuition above what American students can pay and accepting large numbers of international students who can afford full tuition. Nearly half of current doctoral students in STEMM are from other countries. And some state universities in the US are so strapped for cash that most of their STEMM graduate students are foreign-born, primarily hailing from Asia, India, and the Middle East. On top of this, billions of dollars are being spent to educate foreign-born students, *but our immigration laws don't let them stay in the US once they graduate*. Instead, they must return to their home countries, where they often wind up developing technologies that compete with American tech. Tell your legislators to let those we educate in the US stay in the US. As Alberts and Narayanamurti wrote in *Science*, "It is imperative that the United States reconsider its visa and immigration policies, making it easier for foreign students who receive a graduate degree in a STEMM discipline from a U.S. university to receive a green card, while stipulating that each employment-based visa automatically cover a worker's spouse and children." Narayanamurti knows what he's talking about: forty years after immigrating to the US, he became the founding dean of the School of Engineering and Applied Sciences at Harvard University.

••

Hold scientists who won't mentor women accountable. Most grants from the NIH are awarded solely on the basis of research proposed and accomplished. But grants should also take into account the diversity of the trainees in a lab. The NIH should insist that applicants for RO1 grants, the main type of NIH research grant, include a list of all postdoctoral fellows and graduate students ever trained in the lab and what each trainee is doing today. If two grants are found to have equal scientific merit, the lab that trains more diverse and successful trainees should get the grant; these trainee lists should be publicly available. This was Stanford neuroscientist Ben Barres's idea. As Barres noted, if the best male scientists refuse to mentor the best young women, it's basically impossible for women to get ahead.

••

Government agencies and universities currently allow male faculty to use taxpayer-funded research to start all-male tech companies—even in biotech fields where women researchers are leaders. Universities have offices that license the discoveries of faculty and students to begin start-ups. The university should review start-up founders and scientific boards. At the very least, start-ups boards should include leading women researchers in the field. This would help the start-ups, too, since studies report that diversity boosts success and raises profits.

• •

Institutionalize your reforms. Make them last.

The annual supercomputing conference called SC attracts twelve thousand attendees, most of them male. But the six hundred volunteers on its organizing committee had never collected their own demographics. John West of the Texas Advanced Computer Center, where the world's fastest academic supercomputer is located, wanted to do so. "But one year is not enough to institutionalize something in an all-volunteer committee," West pointed out. "The chair of the conference serves for only one year so you can start something that's wonderful, but if the next two chairs haven't bought into the idea, it dies immediately." So after West became chair of the 2016 conference, he convinced the next three chairs to be elected to say, "Yes, we want to do a count of the demographics too." Which they did—and then used this information to increase the number of women involved.

• •

Think beyond academic science. Mentoring PhD students—sixty-two of them—has been one of the highlights of my life. Many have become university professors, and four have been elected to the National Academy of Sciences. But a PhD in science is like a law degree: you can do a lot of different things with it. Many of my former students work in government laboratories and agencies or as university administrators, entrepreneurs, environmentalists, or medical researchers; I also count a venture capitalist, a winemaker, and an artist among them.

• •

Women today hold or have held four of the most prestigious science jobs in the United States: the directorships of the National Science Foundation, the National Science Board, the National Academy of Sciences, and the American Association for the Advancement of Science. But wait . . . something's missing. Those women are all physical scientists: two are geophysicists, one is an astrophysicist, and the fourth is a computer scientist. The majority of women in science today are in the biological sciences, perhaps the most exciting and intellectually challenging field, and yet not one of them is in those leadership positions.

• •

"Tell male teaching assistants and male graduate students that civility and respect toward all students, including women students, is expected, and that demonstrated absence of civility and respect is cause for dismissal. This may be particularly important for foreign male students from male-dominated societies."

—Physicist Fay Ajzenberg-Selove

• •

Know that sexual harassment training can boomerang. People who are told that others are biased can wind up expressing more stereotypes, be less willing to work with others, and treat others in more biased ways. Much of the sexual harassment training today is offered via online mini courses or short videos. Instead, demand programs with qualified live trainers who offer specific examples of inappropriate conduct. Training should establish standards of behavior, instead of trying to influence attitudes and beliefs.

• •

Ban all-male search committees, especially when faculty positions are being filled. Overwhelming evidence shows that they favor hiring far more men than equally qualified women. Unfortunately, many departments today will approve anyone the search committee recommends, without question and without higher-level review.

••

I've seen great progress for women in science over the course of my career, but there is still much more to be done. Leaders in the educational community are beginning to take action.

••

More than a hundred scientific organizations have banded together to combat sexual harassment through the Societies Consortium on Sexual Harassment in STEMM, founded in late 2018 by the American Geophysical Union, the American Association for the Advancement of Science, and the Association of American Medical Colleges. The consortium was formed after a series of high-profile cases alleging sexual harassment in the sciences, as well as the publication of the National Academies' report, which concluded that "organizational climate is, by far, the greatest predictor of the occurrence of sexual harassment." Even the perception that sexual harassment is tolerated increases the likelihood it will occur. The consortium plans to provide model policies for professional, academic, medical, and research institutions to adopt to combat sexual harassment.

The Action Collaborative on Preventing Sexual Harassment in Higher Education is another effort by sixty-three organizations to implement the National Academies' report recommendations.

••

Looking over the list of actions women scientists and their allies still need to take to make the profession truly equitable, I recognize that the years ahead do not look simple or easy. But STEMM professions have improved for women in the four and a half decades since biologist Peter Medawar first wrote "that the world is now such a complicated and rapidly changing place that it cannot even be kept going (let alone improved . . .) without using the intelligence and skill of approximately 50 percent of the human race." The advances are thanks to the women who were brave enough to lead in centuries past: the women who formed the suffrage movement; those who formed sisterhoods and launched the women's rights movement and its recent renaissance; and those who bravely brought about the #MeToo movement and its many manifestations. I hope that, after reading this book, you will better appreciate how women in science have been part of that rich history. We are in a century of breakthroughs in science, technology, and engineering. There is so much left to do—but also so much that we *can* do. If we are all in this together, the future is limitless.

Acknowledgments

A lifetime displayed candidly resembles a garden . . . filled with lovely colorful blossoms . . . dispersed among hidden weeds of time, some of which were perfume-scented and others prickly and occasionally bearing thorns. My very own memory garden is rich with caring and many kindnesses . . . and a few interspersed bristles. Let me acknowledge first those who gave me joy and happiness: my father, Louis Rossi, whose commitment to fairness, equity, and the power of learning provided a foundation and powerful protection; and my sisters, Marie George, Yolanda Frederikse, and Paola Biola, always there to bolster the spirit and provide unconditional love; my beloved husband, Jack Colwell, with whom I shared a lifetime of adventure and exploration; my children, Alison Colwell and Stacie Colwell, both beautiful and talented, who made my life meaningful; and my sons-in-law, Richard Canning and Bruce Ponman—both added a new dimension that is forever deeply appreciated.

My patient assistant, Vickie Lord, suffered with me through endless revisions and anguished searches for precision and context.

My many students, every one of whom is sincerely appreciated, especially those who contributed to the book through their excellent research work and in many discussions—Anwarul Huq, James Kaper, Jody Deming, Tamar Barkay, Ronald Sizemore, James Oliver, John Schwarz, Tatsuo Kaneko, Ronald Citarella, Sharon Berk, Kazuhiro

Kogure, Dawn Allen-Austin, Brian Austin, Ivor Knight, Darlene Roszak MacDonell, Steven Orndorff, Charles Somerville, Paul Tabor, Constance Cepko, Antarpreet S. Jutla, and Nur Hasan.

Many friends and colleagues also contributed to the book. Special thanks go to colleagues Ben Shneiderman, Sam Joseph, Ben Cavari, Carla Pruzzo, Monique Pommepuy, Dominique Hervio-Heath, Patrick Monfort, Nancy Hopkins, Robert Birgeneau, Paul G. Gaffney II, and many other wonderful colleagues, former students, postdoctoral fellows, and visiting scientists from the United Kingdom, Europe, Asia, Latin America, Africa, Australia, New Zealand, and Canada.

I must especially acknowledge my coauthor, Sharon Bertsch McGrayne, who has an uncanny knack and amazing talent for ferreting out facts and data . . . and making sense of it all. Without her collaboration, this book could not have been written.

Special thanks to Priscilla Painton, Megan Hogan, and Susan Rabiner for their patience and goodwill in dealing with emergencies, exigencies, and occasional crises, always with calm and reason.

—RITA ROSSI COLWELL

• •

First, I want to thank Rita Colwell for the opportunity to work with and learn from her. I must also thank Priscilla Painton for her faith in the project and her support for it throughout the process. She was everything I could have hoped for in an editor. Her very able assistant, Megan Hogan, helped in shaping the manuscript, and I thank her for her wonderful work. Victoria Lord of the University of Maryland was an ever-ready source of information and good cheer. As for our agent, Susan Rabiner, this book would not exist without her years of support and perceptive advice.

I also thank the following people who graciously allowed me to interview them.

prologue

Margaret Walsh Rossiter

chapter one: **No, Girls Can't Do That!**

Paola Biola, Barbara Cohn Younger, Jack H. Colwell, Andrew G. DeRocco, John G. Holt, David M. Hovde, Margaret Jauron Luby, Dale Kaiser, Henry Koffler, Lorna Laroe Lieberman, Marianne Mayer Wentzel, Laura L. Mays Hoopes, Harry Morrison, Wenham Museum, Yolanda Rossi Frederikse, Marie Rossi George, Marilyn Treacy Miller Fishman, and William Wiebe

chapter two: **Alone: A Patchwork Education**

Artrice F. Valentine Bader, Anna W. Berkovitz, William J. Browning, George Chapman, Kenneth K. Chew, J. Alfred Chiscon, Martha O. Chiscon, Lorraine Daston, Jody Deming, Bruce Gochnauer, Karen Gochnauer, Nancy Gochnauer, Benjamin D. Hall, Margaret A. Hall, Estella Leopold, John Liston, Richard Y. Morita, Harry A. Morrison, M. Patricia Morse, Eugene Ernest Nester, Father Joseph Allen Panuska, Helen Remick, James T. Staley, Frieda B. Taub, Shirley M. Tilghman, Suzanne E. Vandenbosch, Arthur H. Whiteley, and archivists John D. Bolcer and Kathleen Brennan

chapter three: **It Takes a Sisterhood**

Ben A. Barres, Joan W. Bennett, Sheila Bird, Charlotte G. Borst, Jean E. Brenchley, Lynn Caporale, Eugenie Clark, Carol A. Colgan, Kelly Marjorie M. Cowan, Elizabeth L. R. Donley, Walter R. Dowdle, Richard A. Finkelstein, Angela Ginorio, Michael Goldberg, Florence P. Haseltine, Nancy Hopkins, Rita Horner, Alice S. Huang, Barbara H. Iglewski, Samantha B. Joye, Sally Frost Mason, Anne Morris-Hooke, Frederick C. Neidhardt, Vivian Pinn, Sue V. Rosser, Margaret Walsh Rossiter, Sara Rothman, Karla Shepard Rubinger, Moselio Schaechter, Pat Schroeder, and ASM archivist Jeff Karr. I am also indebted to ASM for a travel grant to visit the Center for the History of Microbiology/ASM Archives (CHOMA), University of Maryland, Baltimore County, April 29–30, 2014.

chapter four: **The Power of Sunlight**

Lotte Bailyn, Penny Chisholm, Nancy Hopkins, Marc A. Kastner, Marcia McNutt, Christiane Nüsslein-Volhard, Shirley Tilghman, and information from Lydia J. Snover, director of institutional research, Office of the Provost, MIT

chapter five: **Cholera**

Amanda Allen, Tamar Barkay, Sharon Berk, Jack H. Colwell, Jody W. Deming, William "Buck" Greenough, D. Jay Grimes, Patricia Guerry, John G. Holt, Anwar Huq, Samuel W. Joseph, Antarpreet S. Jutla, James B. Kaper, Ivor T. Knight, Darlene Roszak MacDonell, Patricia Morse, James D. Oliver, Steve A. Orndorff, Estelle Russek-Cohen, R. Bradley Sack, Ronald Sizemore, Charles Somerville, Mark Stromm, Paul S. Tabor, Michelle L. Trombley, and archivists Leslie Fields, Mount Holyoke College; Louise S. Sherby and Christopher Browne, Hunter College; James Stimpert, Johns Hopkins University; and Jeanne d'Agostino, Memorial Sloan Kettering Cancer Center

chapter six: **More Women = Better Science**

Ruzena Bajcsy, Barry Barish, Diana Bilimoria, Amy Bix, Joseph Bordogna, Norman Bradburn, Norma Brinkley, Mary E. Clutter, Jack H. Colwell, Thomas N. Cooley, Robert W. Corell, Margo H. Edwards, Karl A. Erb, Rachel A. Foster, Jillian Freese, Valerie G. Hardcastle, Alice Hogan, Stacie Furst Holloway, Anwar Huq, Bethany Jenkins, Margaret Leinen, Marcia K. McNutt, Barbara A. Mikulski, Constance A. Morella, Anna Ruth Robuck, Marty Rosenberg, Vernon D. Ross, Tatiana Rynearson, Barbara E. Silver, Howard J. Silver, Alexa Sterling, Philippe M. Tondeur, Michael S. Turner, and NSF archivist Leo Slater

chapter seven: **The Anthrax Letters**

Teresa G. "Terry" Abshire, Bruce Budowle, Thomas A. Cebula, Richard Danzig, R. Scott Decker, Daniel Drell, John Ezzell, Steve Fienberg, Claire M. Fraser, Gigi Kwik Gronvall, Jeanne Guillemin, Paul J. Jackson, Norman Kahn, Paul S. Keim, Michael R. Kuhlman, Vahid Majidi,

Matthew Meselson, Ari A. N. Patrinos, John R. Phillips, Adam Phillippy, Mihai Pop, Jacques Ravel, Timothy D. Read, Steven L. Salzberg, Scott T. Stanley, Ronald A. Walters, David Willman, Linda Zall, and Raymond Zilinskas

chapter eight: **From Old Boys' Clubs to Young Boys' Clubs to Philanthropists**

Linda Babcock, Candida Brush, Alan H. DeCherney, Samantha Joye, Marc Kastner, Robbie Melton, Charles "Chuck" Miller, Amy Millman, Carol A. Nacy, Ginny Orndorff, Steve Orndorff, and Kurt Soderlund

chapter nine: **It's Not Personal—It's the System**

Arturo Casadevall, Jennifer T. Chayes, Jo Handelsman, Nicole Smith, Claude M. Steele, Lee J. Tune, and Jay Van Bavel

chapter ten: **We Can Do It**

Tamar Barkay, Ben Barres, Ruth Ann Bertsch, Constance L. Cepko, Andrei Cimpian, Lorraine Daston, Florence Haseltine, Marie Hicks, Nancy Hopkins, Alice Huang, Crystal H. Johnson, Marcia McNutt, Amy Millman, Lucy Sanders, Shirley M. Tilghman, and John West

I thank them all.

—**SHARON BERTSCH McGRAYNE**

• •

We also want to thank the many people who assisted us or critiqued the book or portions of it in draft form. They include: Theresa G. Abshire, Ashley Bear, Joan Bennett, Fred Bertsch, Ruth Ann Bertsch, Marie-Joelle Dominioni Blaizot, Bruce Budowle, Richard Canning, Davis J. Cassel, Barbara Cohn Younger, Jean Colley, Alison Colwell, Jack H. Colwell, Stacie Colwell, France Córdova, Scott Decker, Jody Deming, Jennifer Doudna, Elizabeth Morrow Edwards, Claire M. Fraser, Yolanda Rossi Frederikse, Marie Rossi George, Angela Ginorio, Maria Y. Giovanni, Jeanne Guillemin, Jo Handelsman, Dominique Hervio-Heath, Alice Huang, Sam Joseph, Antarpreet S. Jutla, Norman

Kahn, Carolyn W. Keating, Ivor Knight, Michael R. Kuhlman, Neal Lane, Hilary Lapin-Scott, Margaret Leinen, Rachel Levinson, Victoria Lord, Gary Machlis, Matthew Meselson, Patrick Monfort, Anne Morris-Hooke, Emilia Muller-Ginorio, James D. Oliver, Ari A. N. Patrinos, John R. Phillips, Monique Pommepuy, Bruce Ponman, Carla Pruzzo, Jacques Ravel, Kitsy Rigler, Debra Samson, Ben Schneiderman, Frieda Taub, Ronald A. Walters, Audrey Weitkamp, David Willman, and Chuck Wilson.

Recommended Reading

Epstein, Paul R., and Dan Ferber. *Changing Planet, Changing Health: How the Climate Crisis Threatens Our Health and What We Can Do About It.* Berkeley: University of California Press, 2011.

Manning, Kenneth R. *Black Apollo of Science: The Life of Ernest Everett Just*. New York: Oxford University Press, 1983.

McGrayne, Sharon Bertsch. *Nobel Prize Women in Science: Their Lives, Struggles, and Momentous Discoveries*. 2nd ed. Washington, DC: National Academies Press, 1998.

National Academies of Sciences, Engineering, and Medicine. *Sexual Harassment of Women: Climate, Culture, and Consequences in Academic Sciences, Engineering, and Medicine.* Washington, DC: National Academies Press, 2018.

National Academy of Sciences. *Seeking Solutions: Maximizing American Talent by Advancing Women of Color in Academia*. Washington, DC: National Academies Press, 2013.

Rosenberg, Charles A. *The Cholera Years: The United States in 1832, 1849, and 1866*. 2nd ed. Chicago: University of Chicago Press, 1987.

Rossiter, Margaret W. *Women Scientists in America.* 3 vols. Baltimore: Johns Hopkins University Press, 1995.

Schwartz, Neena B. *A Lab of My Own.* New York: Rodopi, 2010.

Steele, Claude M. *Whistling Vivaldi: How Stereotypes Affect Us and What We Can Do.* New York: W. W. Norton, 2010.

Willman, David. *The Ames Strain: The Mystery Behind America's Most Deadly Bioterror Attack*. 2nd ed. Brooklyn, NY: February Books, 2014.

Notes

All interviews listed were conducted by phone by Sharon Bertsch McGrayne, unless noted otherwise. Emails were also to McGrayne.

prologue: **Hidden No More**

xiii *Margaret Walsh Rossiter*: Margaret Walsh Rossiter interview, December 8, 2019, and emails from October 6 and December 3 and 4, 2019; Susan Dominus, "Women Scientists Were Written Out of History," *Smithsonian Magazine*, October 2019.

chapter one: **No, Girls Can't Do That!**

1 *Even colleagues can feel intimidated*: Al Chiscon, interview, October 2, 2013.

2 *uncomfortable time to be Italian*: Helene Stapinski, "When America Barred Italians," *New York Times*, June 2, 2017; Rita James Simon, *In the Golden Land: A Century of Russian and Soviet Jewish Immigration in America* (Westport, CT: Praeger, 1997), 15–16; *The Ambivalent Welcome* (Santa Barbara, CA: Praeger, 1993), 72; Mahzarin R. Banaji and Anthony G. Greenwald, *Blindspot: Hidden Biases of Good People* (New York: Delacorte Press, 2013), 77.

2 *knock at the front door*: Marie Rossi George, interview, January 5, 2019.

6 *Today I believe Dr. Box*: In Russell L. Cecil, ed., *A Textbook of Medicine*, 7th ed. (Philadelphia: W. B. Saunders, 1947–48): Cary Eggleston, "Myocardial Infarction," 1124–25; Dickinson W. Richards, "Diseases of the Bronchi," 917–18, 920–23; Russell L. Cecil, "Pneumococcal Pneumonia," 129.

6 *high school yearbook*: Barbara Cohen Younger, interview, July 15, 2013; Lorna Laroe Lieberman, interview, July 16, 2013.

7 *My chemistry teacher*: H. S. Gutowsky, *1972–3 Annual Report, School of Chemical Sciences* (Champaign: University of Illinois Champaign-Urbana, August 1973).

7 *Astronomer Nancy Roman*: Emily Langer, "Nancy Grace Roman, Astronomer, Celebrated as 'Mother' of the Hubble, Dies at 93," *Washington Post*, December 28, 2018.

8 *It's no wonder*: "President Kennedy's Commission on the Status of Women," Washington, DC: PCSW, 1961; and 60.

8 *Lamont, Harvard's undergraduate library*: "HUC Demands Cliffies Be Kept Out of Lamont," *Harvard Crimson*, January 4, 1966.

8 *The federal government was transforming*: Robert W. Topping, *A Century & Beyond: The History of Purdue* (Lafayette, IN: Purdue University Press, 1989), 255, 264.

10 *golden years of government support*: Margaret W. Rossiter, *Women Scientists in America: Before Affirmative Action, 1940–1972*, vol. 2 (Baltimore: Johns Hopkins University Press, 1995), 48–49, 123.

10 *Laura L. Mays Hoopes*: Laura L. Mays Hoopes, *Breaking Through the Spiral Ceiling: An American Woman Becomes a DNA Scientist* (Morrisville, NC: Lulu Publishing, 2013), 29–34.

11 *Gerty Radnitz Cori*: Sharon Bertsch McGrayne, *Nobel Prize Women in Science: Their Lives, Struggles, and Momentous Discoveries* (Washington, DC: National Academies Press, 1998), 93–116.

11 *Maria Goeppert Mayer*: McGrayne, *Nobel Prize Women in Science*, 175–200.

11 *Yvonne Brill*: Douglas Martin, "Yvonne Brill, a Pioneering Rocket Scientist, Dies at 88," *New York Times*, March 30, 2013; Margaret Sullivan, "Gender Questions Arise in Obituary of Rocket Scientist and Her Beef Stroganoff," *New York Times*, March 30, 2013.

12 *Barbara McClintock*: McGrayne, *Nobel Prize Women in Science*, 144–74.

12 *Rosalind Franklin*: McGrayne, *Nobel Prize Women in Science*, 304–32; Brenda Maddox, *Rosalind Franklin: The Dark Lady of DNA* (New York: HarperCollins, 2002); James O. Watson, *Genes, Girls, and Gamow* (Oxford, UK: Oxford University Press, 2001); Charlotte Hunt-Grubbe, "The Elementary DNA of Dr. Watson," *Sunday Times* (London), October 14, 2007.

12 *Watson would be stripped*: Adam Rutherford, "He May Have Unravelled DNA, but James Watson Deserves to Be Shunned," *Guardian* (UK), December 1, 2014.

13 *Alice Catherine Evans*: Rita R. Colwell, "Alice C. Evans: Breaking Barriers," *Yale Journal of Biology and Medicine* 72, no. 5 (September–October 1999).

14 *twenty-nine female full professors*: Rossiter, *Women Scientists in America*, 2:128–29.

14 *Dorothy Powelson*: University of Georgia yearbook for 1937 and 1938, student records, and Alumni Association; University of Wisconsin Graduate School 1942–1945, faculty employment form, Sigma Delta Epsilon scrapbook; University of Maine in Orono, "Workers in Land Grant Colleges and Stations," US Department of Agriculture Miscellaneous Publication 677, Washington, DC, April 1949, 42; July 1959 photograph, thanks to Purdue archivist Stephanie Schmidt and her scientific background and Purdue positions; *Purdue University Bulletin School of Science Announcements* for the years 1952–53, 1953–54, 1954–55, 1955–56, 1956–57, 1957–58, and 1958–59, Purdue University, Lafayette, IN.

15 *I didn't really like*: John G. Holt, interview, November 1, 2012; Chiscon, interview.

15 *two Stanford University men*: Dale Kaiser and Martin Dworkin, "From Glycerol to the Genome," in *Myxobacteria*, ed. David E. Whitworth (Washington, DC: ASMScience, 2008), 3–15.

16 *but didn't deny*: Henry Koffler, interview, January 2, 2013.

chapter two: **Alone: A Patchwork Education**

18 *The system was casual*: David Stadler, "Herschel Roman and 50 Years of Genetics at the University of Washington" (presentation to the University of Washington Department of Genetics seminar, Seattle, WA, December 14, 1992); Nancy Hopkins, "The Changing Status and Number of Women in Science and Engineering at MIT" (keynote to the MIT150 symposium Leaders in Science and Engineering: The Women of MIT, MIT, Cambridge, MA, March 28, 2011), 4.

18 *scale in the 1950s and '60s*: Linda Eisenmann, *Higher Education for Women in Postwar America 1945–1965* (Baltimore, MD: Johns Hopkins University Press, 2006); Lorraine Daston, interview, Seattle, WA, April 19, 2017.

19 *Margaret A. Hall's history of women*: Margaret A. Hall, "A History of Women Faculty at the University of Washington, 1896–1970" (PhD diss., University of Washington, 1984), and interview, Bellevue, WA, April 11, 2013; Benjamin D. Hall, interview, Bellevue, WA, April 11, 2013.

20 *Erna Gunther*: Viola E. Garfield and Pamela T. Amoss, "Erna Gunther (1896–1982)," *American Anthropologist* 86, no. 2 (June 1984).

20 *Dixy Lee Ray*: Erik Ellis, "Dixy Lee Ray, Marine Biology, and the Public Understanding of Science in the United States (1930–1970)" (PhD thesis, Oregon State University, 2006); Dixy Lee Ray with Lou Guzzo, *Trashing the Planet: How Science Can Help Us Deal with Acid Rain, Depletion of the*

Ozone, and Nuclear Waste (Among Other Things) (Washington, DC: Regnery Gateway, 1990).

20 *Dora Priaulx Henry*: P. A. McLaughlin, "Dora Priaulx Henry (24 May 1904–16 June 1999)," *Journal of Crustacean Biology* 20, no. 1 (2000).

20 *Helen Riaboff Whiteley*: Arthur H. Whiteley, interview, Seattle, WA, January 9, 2013; Eric Pryne, "Helen R. Whiteley, UW Professor," *Seattle Times*, January 1, 1991; Laura L. Mays Hoopes, interview, April 8, 2015.

20 *the state's anti-nepotism laws and university regulations*: Hall, *A History of Women Faculty*; Rossiter, *Women Scientists in America*, 2:123–28, 138–41.

21 *Frieda B. Taub*: Frieda B. Taub, interview, Seattle, WA, December 15, 2012, and emails.

23 *Anna Whitehouse Berkovitz*: Anna W. Berkovitz, interview, April 7, 2014.

24 *Violet Bushwick Haas*: Pamela G. Coxson, "In Remembrance of Violet Bushwick Haas (1926–1986)," *AWM Newsletter* 16, no. 4 (1986): 2–3.

25 *Chiscon*: Al Chiscon and Martha Chiscon, interviews, October 2, 2013.

26 *John Liston*: John Liston, interviews, Bothell, WA, May 31 and August 9, 2012.

28 *Emmy Klieneberger-Nobel*: Emmy Klieneberg-Nobel, *Memoirs* (London: Academic Press, 1980), 80–81.

29 *the size of three large refrigerators*: Howard Wainer and Sam Savage, review of *The Theory That Would Not Die*, by Sharon Bertsch McGrayne, *Journal of Educational Measurement* 49, no. 2 (June 25, 2012).

29 *The prestigious journal* Nature: R. R. Colwell and J. Liston, "Taxonomy of Xanthomonas and Pseudomonas," *Nature* 191 (1961): 617–19.

31 *Shirley M. Tilghman*: Tilghman Presidential Speeches, Princeton University, February 28, 2006.

32 *Margaret Briggs Gochnauer*: Nancy, Bruce, and Karen Gochnauer, interviews, December 2012.

32 *Elizabeth McCoy*: Rossiter, *Women Scientists in America*, 2:134, table 6.3.

34 *small notice pinned to a bulletin board*: Richard Y. Morita, interview, 2013.

34 *George Chapman*: George Chapman, interviews, August 13 and 22, 2012, and letter, August 29, 2012.

34 *Sarah P. Gibbs*: Sarah P. Gibbs, "Fighting for My Own Agenda: A Life in Science," in *Our Own Agendas: Autobiographical Essays by Women Associated with McGill University*, eds. Margaret Gillett and Ann Beer (Montreal: McGill–Queen's University Press, 1995).

35 *a (male) West Coast competitor*: Confidential interview, August 30, 2016.

37 *Nearby Johns Hopkins University had only two women*: Rossiter, *Women Scientists in America*, 2:134, table 6.3.

38 *Artrice Valentine Bader*: Artrice F. Valentine Bader, interviews, March 5 and 20, 2018.

38 *next year:* Chapman, interviews and letter; Father Joseph Allen Panuska, interview, August 18, 2010.

chapter three: **It Takes a Sisterhood**

40 *Bernice R. "Bunny" Sandler*: Bernie "Bunny" Sandler, " 'Too Strong for a Woman'—The Five Words That Created Title IX," *Equity & Excellence in Education* 33, no. 1 (April 2000), and "Title IX: How We Got It and What a Difference It Made," *Cleveland State Law Review* 55, no. 473 (2007); www.bernicesandler.com; Rossiter, *Women Scientists in America*, 2:374–77, and vol. 3, *Forging a New World Since 1972*, xvii–xviii, 21–22.

42 *Jonathan Spivak of the* Wall Street Journal: Jonathan Spivak, "New Higher-Education Bill Provides More Funds, but Sex-Bias Section Could Spark Controversy," *Wall Street Journal*, July 13, 1972.

42 *Thousands of women suddenly received big pay raises*: Sandler, "Title IX," 486.

42 *educational institutions had to publicly advertise*: Rossiter, *Women Scientists in America*, 3:26, 31.

43 *Ben A. Barres*: Ben A. Barres, emails to McGrayne, September 11, 16, 18, and 19, 2016. Also Barres, "Does Gender Matter?" *Nature* 442 (July 13, 2006), "Some Reflections on the 'Dearth' of Women in Science" (lecture, Harvard University, March 17, 2008), letter to the editor, *New York Times*, August 12, 2017, and *The Autobiography of a Transgender Scientist* (Cambridge, MA: MIT Press, 2018). Also about Barres: Shankar Vedantam, "Male Scientist Writes of Life as Female Scientist," *Washington Post*, July 13, 2006; Amy Adams, "Barres Examines Gender, Science Debate and Offers a Novel Critique," *Stanford Report*, July 26, 2006; Niuniu Teo, "Transgender Professor Advocates for Women in Science," *Stanford Daily*, October 4, 2013; Neil Genzlinger, "Ben Barres, Neuroscientist and Equal-Opportunity Advocate, Dies at 63," *New York Times*, December 29, 2017; Matt Schudel, "Ben Barres, Transgender Brain Researcher and Advocate of Diversity in Science, Dies at 63," *Washington Post*, January 2, 2018.

45 *Age of Revolving Doors*: Rossiter, *Women Scientists in America*, 3:33.

45 *Sally Frost Mason*: Sally Frost Mason, interview, April 8, 2014, and email, April 16, 2014.

45 *Lynn Caporale*: Lynn Caporale, interview, October 15, 2015.

46 *Rita Horner*: Rita Horner, interview, December 19, 2012.

46 *Sue V. Rosser*: Sue V. Rosser, *Breaking Into the Lab: Engineering Progress for Women in Science* (New York: New York University Press, 2012), 1.

47 *the number of women on prestigious faculties actually dropped*: Rossiter, *Women Scientists in America*, 3:33.

47 *the Shark Lady*: Eugenie Clark, *Lady with a Spear* (New York: Harper & Row, 1951), *The Lady and the Sharks* (New York: Harper & Row, 1969), and email, January 24, 2013.

48 *Women in the Endocrine Society worked hard . . . Yalow*: Neena B. Schwartz, *A Lab of My Own* (New York: Rodopi, 2010), 118.

49 *Huang was born in China*: Huang, interviews, April 13, 2014, and May 17, 2019.

50 *Divorce or widowhood could end a woman's career*: Katy Steinmetz, "Esther Lederberg and Her Husband Were Both Trailblazing Scientists and Why More People Heard of Him," *Time*, April 11, 2019; Rossiter, *Women Scientists in America*, 2:118, 155, 331; *American Men and Women of Science 1992–1993*, 18th ed. (New Providence, NJ: R. R. Bowker, 1992).

50 *an eminent Canadian professor of psychology*: Hans Selye, "Who Should Do Research," in *From Dream to Discovery: On Being a Scientist* (New York: McGraw-Hill, 1964).

50 *study of the problems facing female PhD biologists*: Eva Ruth Kashket et al., "Status of Women Microbiologists: A Preliminary Report," *Science* 183, no. 4124 (February 8, 1974): 488–94; Mary Louise Robbins, "Status of Women Microbiologists: A Preliminary Report," *ASM News* 37 (1971): 34–40.

51 *giant graph posted in a hallway*: "Committee on Women Enters Second Quarter Century," *ASM News* 63, no. 3 (1996): 148–89.

51 *Mary "Polly" Bunting*: Huang, interview April 13, 2014; Linda Eisenmann, *Higher Education for Women in Postwar America, 1945–1965* (Baltimore: Johns Hopkins University Press, 2006), 179–205; Karen W. Arenson, "Mary Bunting-Smith, Ex-President of Radcliffe, Dies at 87," *New York Times*, January 23, 1998.

53 *hematologist Judith Graham Pool and endocrinologist Neena B. Schwartz*: Schwartz, *A Lab of My Own*, 110; Lisa Lepson, "Judith Graham Pool 1919–1975," *Jewish Women: A Comprehensive Historical Encyclopedia*, Jewish Women's Archive, February 27, 2009, https://jwa.org/encyclopedia/article/pool-judith-graham.

53 *Schwartz had been hired*: Schwartz, *A Lab of My Own*, 71.

53 *women in the medical sciences weren't getting*: Rossiter, *Women Scientists in America*, 3:3–4; Schwartz, *A Lab of My Own*, 110–11.

53 *Bunny Sandler and AWIS*: Rossiter, *Women Scientists in America*, 3:29–40.

54 *Marjorie Crandall*: Crandall and Louden letter, in Colwell: Presidential: Boards (PSAB): folder 20, ASM Archives, Center for the History of Microbiology/ASM Archives (CHOMA), University of Maryland, Baltimore County.

56 *750 "expert" volunteers*: Jeff Karr (ASM archivist), email, July 9, 2014.

56 *films directed by men*: Sara Buckley, "Women Make Gains in Independent Films," *New York Times*, June 19, 2019.

56 *Meeting in the home of Sara W. Rothman*: Sara Rothman, interview, June 19, 2014.

57 *Albert Balows*: Letter to Frederick Neidhardt, June 14, 1982, ASM Archives.

57 *Frederick C. Neidhardt*: Frederick C. Neidhardt, interview, May 2014, and "Status of Women in ASM," *ASM Newsletter* (October 1982), ASM Archives.

58 *Anne Morris-Hooke*: Anne Morris-Hooke, interview, May 5, 2014.

59 *Jean E. Brenchley*: Jean E. Brenchley, interview, April 18, 2014.

60 *about his "concern"*: John C. Sherris, letter to Viola Mae Young-Horbath, July 6, 1984, in Colwell: Presidential: Boards (PSAB): folder 20, ASM Archives.

60 *Viola Mae Young-Horvath*: Viola Mae Young-Horvath, letter to John C. Sherris, July 20, 1984, in Colwell: Presidential: Boards (PSAB): folder 20, ASM Archives.

60 *Outgoing president Moselio Schaechter*: Morris-Hooke, interview; Rothman, interview.

60 *Tempers cooled slowly*: Barbara H. Iglewski, interview, August 14, 2014.

62 *Consider the case of Samantha "Mandy" Joye*: Samantha Joye, interview, September 4, 2015.

62 *Consider this story, too*: Kelly Marjorie M. Cowan, interview, May 25, 2017.

chapter four: The Power of Sunlight

63 *"There stood a scientist I didn't know 'personally'"*: Nancy Hopkins, "The Changing Status and Number of Women in Science and Engineering at MIT" (keynote to the MIT150 symposium Leaders in Science and Engineering: The Women of MIT, MIT, Cambridge, MA, March 28, 2011), and afterword to chapter 8 of *Becoming MIT: Moments of Decision*, ed. David Kaiser (Cambridge, MA: MIT Press, 2010), 187–92.

63 *Hopkins sometimes regrets having shared this story*: Nancy Hopkins, interview with Colwell and McGrayne, January 25, 2018.

64 *The story . . . starts when*: Hopkins, "The Changing Status," 5, 7, 13.

64 *I had my own opportunity to observe*: "Report of the Visiting Team to the Commission on Institutions of Higher Education of the New England Association of Schools and Colleges on the Subject of the Educational Program at Massachusetts Institute of Technology," April 8–11, 1979. Submitted by Visiting Team Chairman, in the records of the Office of the Provost (AC-0007), box 44, folder: New England Association of Schools and Colleges.

65 *Hopkins was not yet a feminist*: Hopkins, "The Changing Status," 9–10ff.

66 *any lack of success*: Nancy Hopkins, "Reflecting on Fifty Years of Progress for Women in Science," *DNA and Cell Biology* 34, no. 3 (2015).

66 *Christiane Nüsslein-Volhard*: Christiane Nüsslein-Volhard, interview, Tübingen, Germany, April 23, 1998; Sharon Bertsch McGrayne, *Nobel Prize Women in Science*, 2nd ed. (New York: Basic, 1998), 380–408.

66 *Hopkins was interested in the genetics of behavior*: Nancy Hopkins, interview, March 20, 1998.

66 *Hopkins fell in love with zebrafish*: Nancy Hopkins, interview, January 10, 2019; Robin Wilson, "An MIT Professor's Suspicion of Bias Leads to a New Movement for Academic Women," *Chronicle of Higher Education*, December 3, 1999.

67 *When a senior scientist who'd praised reviews she'd written*: Hopkins, interview, January 25, 2018.

67 *A new genetics course she'd developed*: Andrew Lawler, "Tenured Women Battle to Make It Less Lonely at the Top," *Science* 286, no. 5443 (November 12, 1999).

67 *Male undergraduates "would not believe scientific information spoken by a woman"*: Hopkins, interviews, July 10, 2019, and January 25, 2018; Hopkins, "The Changing Status," 10.

67 *Hopkins found herself going to work each day*: Hopkins, interview, January 25, 2018.

67 *The men had bigger labs*: Lawler, "Tenured Women Battle"; MIT Museum, "A Study on the Status of Women Faculty in Science at MIT, 1996–1999," MIT, 2011.

67 *Hopkins was furious*: Wilson, "An MIT Professor's Suspicion of Bias"; Corie Lok, "Nancy Hopkins Named Xconomy's 2018 Lifetime Achievement Award Winner," *Xconomy*, July 24, 2018.

68 *Mary-Lou Pardue*: Hopkins, "The Changing Status," 12–13.

68 *these women were more distinguished*: Hopkins, "The Changing Status," 16, and "Reflecting on Fifty Years of Progress for Women in Science."

68 *they decided to work in secret*: Hopkins, "The Changing Status."

69 *"death by a thousand pinpricks"*: Lawler, "Tenured Women Battle."

69 *For Birgeneau . . . the meeting was "simply overwhelming*: Lawler, "Tenured Women Battle"; Robert Birgeneau, interview with Colwell and McGrayne, June 6, 2019.

69 *His biggest challenge proved to be*: Birgeneau, interview, June 6, 2019.

69 *the data were stark*: Wilson, "An MIT Professor's Suspicion of Bias."

70 *Birgeneau began correcting problems*: Wilson, "An MIT Professor's Suspicion of Bias"; Genevieve Wanucha, "Women in Marine Science Seize the Day," Oceans at MIT, October 9, 2014, http://oceans.mit.edu/featured-stories/women-marine-science.html.

70 *how had such inequities come to exist*: *MIT Faculty Newsletter* Special Edition, XI no. 4, March 1999; Hopkins, interview, January 25, 2018.

71 *women reporters heard that MIT*: Lotte Bailyn, "Putting Gender on the Table," in *Becoming MIT: Moments of Decision*, ed. David Kaiser (Cambridge, MA: MIT Press, 2014); Kate Zernike, "MIT Women Win Fight Against Bias; In a Rare Move, School Admits Discrimination Against Female Professors," *Boston Globe*, March 21, 1999; Carey Goldberg, "MIT Admits Discrimination Against Female Professors," *New York Times*, March 23, 1999.

71 *1983 report compiled by MIT graduate students*: "Barriers to Equality in Academia: Women in Computer Science at MIT; Prepared by Female Graduate Students and Research Staff in the Laboratory of Computer Science and the Artificial Intelligence Lab at MIT," MIT, February 1983.

71 *Hopkins felt her bitterness and pain begin to slip*: Hopkins, interview, January 25, 2018.

72 *A* Wall Street Journal *editorial denounced the MIT report*: Editorial, "Gender Bender," *Wall Street Journal*, December 29, 1999; Bailyn, "Putting Gender on the Table."

72 *Vest and Birgeneau wrote back*: "Vest, Birgeneau, Answer News Critique of MIT Gender Study," *MIT Tech Talk*, January 12, 2000.

72 *a model for reforming*: Bailyn, "Putting Gender on the Table"; Zernike, "Gains, and Drawbacks, for Female Professors"; Hopkins, speaking at the Rosalind Franklin Society annual meeting, December 17, 2014.

73 *Far from being sidelined*: "A Study on the Status of Women Faculty in Science at MIT," 1999, 7.

73 *Hopkins's career took off, too*: Wilson, "An MIT Professor's Suspicion of Bias."

73 *Marcia K. McNutt*: Marcia McNutt, interview at Philosophical Society of Washington, Washington, DC, January 5, 2018.

73 *How did the revolt . . . succeed?*: Bailyn, "Putting Gender on the Table"; Hopkins, interview; Birgeneau, interview; Penny Chisholm, interview, June 18, 2019.

73 *Sallie W. "Penny" Chisholm*: Wanucha, "Women in Marine Science Seize the Day."

74 *Shirley Tilghman*: Shirley Tilghman, interview, May 21, 2019.

74 *"Rule #1: Time alone . . ."*: Hopkins, speaking at the Rosalind Franklin Society annual meeting, December 17, 2014.

74 *"Double Expectation"*: Marcia K. McNutt, interview, Philosophical Society of Washington, Washington, DC, January 5, 2018.

75 *Lotte Bailyn*: Lotte Bailyn, "Putting Gender on the Table."

75 *the price you paid*: Nancy Hopkins, interview with Colwell and McGrayne, March 15, 2018.

75 *MIT's School of Science*: Hopkins, speaking at the Rosalind Franklin Society annual meeting, December 17, 2014; Lydia J. Snover (director, institutional research, Office of the Provost, MIT), email, June 12, 2019.

chapter five: **Cholera**

77 *in 1976, many Bangladeshis did not take girl babies*: Stan D'Souza and Lincoln C. Chen, "Sex Differentials in Mortality in Rural Bangladesh," *Population and Development Review* 66, no. 2 (June 1980); Lincoln Bin Chen et al., "Sex Bias in the Family Allocation of Food & Health Care in Rural Bangladesh," *Population and Development Review* 7, no. 1 (March 1981). We are indebted to Anwar Huq for these references.

77 *a team of public health officers*: Richard A. Cash et al., "Response of Man to Infection with *Vibrio cholerae*. I. Clinical, Serologic, and Bacteriologic Responses to a Known Inoculum," *Journal of Infectious Diseases* 129, no. 1 (January 1, 1974).

78 *I was respected as a scientific administrator*: William "Buck" Greenough, interview, Baltimore, MD, March 21, 2013.

78 *Filippo Pacini*: Christopher Hamlin, *Cholera: The Biography* (Oxford, UK: Oxford University Press, 2009), 9, 160.

78 *John Snow*: Charles E. Rosenberg, *The Cholera Years* (Chicago: University of Chicago, 1962); Hamlin, *Cholera*, 179–208.

79 *Edward Frankland*: S. M. McGrayne, "Clean Water and Edward Frankland," in *Prometheans in the Lab: Chemistry and the Making of the Modern World* (New York: McGraw-Hill, 2001), 43–57.

79 *Robert Koch*: Hamlin, *Cholera,* 209–66.

79 *Fred Singleton*: Fred L. Singleton, "Effects of Temperature and Salinity on *Vibrio cholerae* Growth," *Applied and Environmental Microbiology* 44, no. 5 (December 1982).

80 *Thomas S. Kuhn*: Thomas S. Kuhn, *The Structure of Scientific Revolutions* (Chicago: University of Chicago Press, 1962).

80 *Max Planck*: Max Planck, *Scientific Autobiography and Other Papers*, trans. F. Gaynor (New York: Philosophical Library, 1950), 33–34.

83 *Robert Pollitzer's 1,019-page compendium*: Robert Pollitzer, *Cholera* (Geneva, Switzerland: World Health Organization, 1959).

83 *Frances Adelia Hallock*: Frances Adelia Hallock, "The Coccoid Stage of Vibrios," *Transactions of the American Microscopical Society* 78, no. 2 (April 1959); "Coccoid Stage of *Vibrio comma*," *Transactions of the American Microscopical Society* 79, no. 3 (July 1960); "The Life Cycle of *Vibrio alternans* (sp. nov)," *Transactions of the American Microscopical Society* 79, no. 4 (October 1960). About Hallock: Hallock, 1940 Census, New York

City; Sloan-Kettering Memorial Hospital, "Halter Retires after Third of Century on Hospital Staff," *Fourfront* 4, no. 5 (February 1961); archivists Leslie Fields, Mount Holyoke College; Louise S. Sherby and Christopher Browne, Hunter College; James Stimpert, Johns Hopkins University; and Jeanne d'Agostino, Sloan-Kettering Memorial Hospital.

87 *The conference was held in a hotel*: Carl H. Oppenheimer, ed., *Marine Biology IV: Proceedings of the Fourth International Interdisciplinary Conference: Unresolved Problems in Marine Microbiology* (New York: The New York Academy of Sciences, 1968).

88 *Both a DNA study by my first graduate student*: R. V. Citarella and R. R. Colwell, "Polyphasic Taxonomy of the Genus *Vibrio*: Polynucleotide Sequence Relationships among Selected *Vibrio* Species," *Journal of Bacteriology* 104, no. 1 (October 1970): 434–42.

88 *my computer programs*: R. R. Colwell, "Polyphasic Taxonomy of the Genus *Vibrio*: Numerical Taxonomy of *Vibrio cholerae*, *Vibrio parahaemolyticus*, and Related *Vibrio* Species," *Journal of Bacteriology* 104, no. 1 (October 1970): 410–33.

89 *Harold E. Varmus*: Harold E. Varmus, *The Art and Politics of Science* (New York: W. W. Norton, 2009).

89 *"Finkelstein in his younger days"*: W. E. van Heyningen and John R. Seal, *Cholera: The American Scientific Experience 1947–1980* (Boulder, CO: Westview Press, 1983), 169–71.

89 *Sixty US medical schools still used*: J. Robert Willson, Clayton R. Beecham, and Elsie Reid Carrington, *Gynecology*, 3rd ed. (St. Louis, MO: C. V. Mosby Co., 1973), 47–49.

90 *Medical research itself was so dominated by men*: Florence P. Haseltine, interview, September 25, 2016, and Haseltine, ed., *Women's Health Research: A Medical and Policy Primer* (Washington, DC: American Psychiatric Press, 2005); Pat Schroeder, interview, May 5, 2018, and Schroeder, *24 Years of House Work . . . and the Place Is Still a Mess* (Kansas City, MO: Andrews McMeel, 1998); John A. Kastor, *The National Institutes of Health 1991–2008* (New York: Oxford University Press, 2010); Bernadine Healy, *A New Prescription for Women's Health: Getting the Best Medical Care in a Man's World* (New York: Penguin Books, 1996), 1–27.

91 *a young Japanese man*: Tatsuo Kaneko and R. R. Colwell, "Incidence of *Vibrio parahaemolyticus* in Chesapeake Bay," *Journal of Applied Microbiology* 30, no. 2 (1975): 251–57.

91 *Kaneko provoked an unanticipated budgetary setback*: Elizabeth Shelton, "Bacteria Infect Bay's Seafood," *Washington Post*, August 29, 1970; U.S. Department of the Interior, Bureau of Commercial Fisheries, with Biological Laboratory, Oxford, MD, "Microbiology of Marine and Estuarine

Invertebrates," Contract 14-17-0003-149, January 1, 1966–February 28, 1970, $83,000; and National Oceanic and Atmospheric Administration (Sea Grant Program), "*Vibrio parahaemolyticus* and Related Organisms in Chesapeake Bay—Isolation, Pathogenicity and Ecology," Grant 04-3-158-7, August 15, 1972–August 14, 1974, $116,100.

94 *James B. Kaper*: R. R. Colwell, James B. Kaper, and S. W. Joseph, "*Vibrio cholerae*, *Vibrio parahaemolyticus*, and Other Vibrios: Occurrence and Distribution in Chesapeake Bay," *Science* 198, no. 4315 (October 28, 1977): 394–6.

95 *microbiology was so divided that*: John G. Holt, interview, November 1, 2012.

95 *I was an easy target for jokes*: James B. Kaper, interview, Baltimore, MD, March 19, 2013; Ronald Sizemore, interview, August 8, 2014; Huq, interview, February 7, 2013.

96 *Anwar Huq*: Anwar Huq, interview, February 7, 2013, and Huq et al., "Ecological Relationships between *Vibrio cholerae* and Planktonic Crustacean Copepods," *Applied and Environmental Microbiology* 45, no. 1 (January 1983): 275–83.

97 *Jack was always surprised*: Jack H. Colwell, interviews, Bethesda, MD, April 18 and November 29, 2017.

97 *William "Buck" Greenough*: William "Buck" Greenough, interview, Baltimore, MD, March 21, 2013.

100 *The United States had had almost no cholera cases since 1914*: Kaper, interview, Baltimore, MD, March 19, 2013, and Kaper et al., "Molecular Epidemiology of *Vibrio cholerae* in the U.S. Gulf Coast," *Journal of Clinical Microbiology* 16, no. 1 (1982).

101 *The paper we wrote about those Louisiana crabbers*: Huai Shu Xu et al., "Survival and Viability of Nonculturable *Escherichia coli* and *Vibrio cholerae* in the Estuarine and Marine Environment," *Microbial Ecology* 8, no. 4 (1982): 313–23.

101 *I submitted the paper to a colleague*: Samuel W. Joseph, interview, March 1, 2017.

101 *Darlene Roszak*: Darlene Roszak MacDonell, interviews, August 27, 2015, and May 7, 2017, and Roszak and R. R. Colwell, "Survival Strategies of Bacteria in the Natural Environment," *Microbiological Reviews* 51, no. 3 (September 1987).

102 *A cascade of studies*: Charles Somerville and Ivor T. Knight, interviews, March 6, 2018; I. T. Knight, J. DiRuggiero, and R. R. Colwell, "Direct Detection of Enteropathogenic Bacteria in Estuarine Water Using Nucleic Acid Probes," *Water Scientific Technology* 24, no. 2 (1991): 261–66.

102 *Mike Levine*: R. R. Colwell et al., "Viable but Non-culturable *Vibrio*

cholerae O1 Revert to a Cultivable State in the Human Intestine," *World Journal of Microbiological Biotechnology* 12, no. 1 (1996): 28–31.

102 *more than fifty disease-causing bacterial species*: Jim Oliver, "Healthy Waters, Healthy People: A Tribute to Rita Colwell," *Microbe* 11, no. 4 (May 30, 2015).

103 *I sat down to count*: R. R. Colwell, "Global Climate and Infectious Disease: The Cholera Paradigm," *Science* 274, no. 5295 (December 20, 1996): 2025–31.

103 *Vaccines can be used to eradicate smallpox and polio*: R. R. Colwell, "Cholera and the Environment: A Classic Model for Human Pathogens in the Environment" (speech delivered at the American Association for the Advancement of Science annual meeting, Seattle, WA, February 14, 2004).

104 *hauling water is women's work*: Michelle L. Trombley, "Strategy for Integrating a Gendered Response in Haiti's Cholera Epidemic" (briefing note, UNICEF Haiti Child Protection Section/Gender-Based Violence Program, December 2, 2010).

104 *Richard Cash, a Harvard Public Health researcher*: R. A. Cash et al., "Bacteriologic Responses to a Known Inoculum," *Journal of Infectious Diseases* 129, no. 11 (January 1, 1974).

104 *World Bank funded projects to drill*: M. F. Hossain, "Arsenic Contamination in Bangladesh: An Overview," *Agriculture Ecosystems and Environment* 113, no. 1 (April 2006).

105 *sari sieves*: Anwar Huq et al., "A Simple Filtration Method to Remove Plankton-Associated *Vibrio cholerae* in Raw Water Supplies in Developing Countries," *Applied and Environmental Microbiology* 62, no. 7 (July 1996); Anwar Huq et al., "Simple Sari Cloth Filtration of Water Is Sustainable and Continues to Protect Villagers from Cholera in Matlab, Bangladesh," *mBio* 18, no. 1 (2010).

105 *a devastating earthquake*: Centers for Disease Control and Protection, "Cholera in Haiti," www.cdc.gov/cholera/haiti/index.html, accessed February 14, 2019.

106 *What caused the outbreak*: Nur A. Hasan et al., "Genomic Diversity of 2010 Haitian Cholera Outbreak Strains," *PNAS* 109, no. 29 (July 17, 2012).

106 *nutritionists from Cornell and the University of Virginia*: Jie Liu et al., "Pre-Earthquake Non-Epidemic *Vibrio cholerae* in Haiti," *Journal of Infection in Developing Countries* 8, no. 1 (2014): 120–22.

106 *There isn't just one kind of cholera epidemic*: Antarpreet S. Jutla et al., "Environmental Factors Influencing Epidemic Cholera," *American Journal of Tropical Medicine and Hygiene* 89, no. 3 (September 4, 2013): 597–607.

107 *Our definition of "cholera" may also need to be revised*: Brianna Lindsey

et al., "Diarrheagenic Pathogens in Polymicrobial Infections," *Emerging Infectious Disease* 17, no. 4 (April 2011).

107 *genetic fluidity*: Noémie Matthey et al., "Neighbor Predation Linked to Natural Competence Fosters the Transfer of Large Genomic Regions in *Vibrio cholerae*," *eLife* 8 (2019), DOI 10.7554/eLife.48212; Nik Papageorgiou, "The Cholera Bacterium Can Steal Up to 150 Genes in One Go," EPFL, October 8, 2019, https://actu.epfl.ch/news/the-cholera-bacterium-can-steal-up-to-150-genes—3/.

107 *three out of four countries*: World Health Organization, "Number of Reported Cholera Cases," Global Health Observatory data, 2019.

107 *With global warming, vibrios and vibrio-caused diseases*: Luigi Vezzulli et al., "Climate Influence on Vibrio and Associated Human Diseases during the Past Half-Century in the Coastal North Atlantic," *PNAS* 113, no. 34 (August 23, 2016).

108 *When NASA's Landsat satellites started collecting data*: Brad Lobitz et al., "Climate and Infectious Disease: Use of Remote Sensing for Detection of *Vibrio cholerae* by Indirect Measurement," *PNAS* 97, no. 4 (February 15, 2000).

109 *Guillaume Constantin de Magny*: Guillaume Constantin de Magny et al., "Environmental Signatures Associated with Cholera Epidemics," *PNAS* 105, no. 46 (November 18, 2008); Timothy E. Ford et al., "Using Satellite Images of Environmental Changes to Predict Infectious Disease Outbreaks," *Emerging Infectious Diseases* 15, no. 9 (2009).

110 *Dividing Yemen into regions the size*: Steve Cole, "NASA Investment in Cholera Forecasts Helps Save Lives in Yemen," NASA, press release, August 27, 2018, https://www.nasa.gov/press-release/nasa-investment-in-cholera-forecasts-helps-save-lives-in-yemen; Civil Service Awards 2019 (UK), "Fergus McBean," civilserviceawards.com/award-nominee/Fergus-McBean.

chapter six: More Women = Better Science

111 *On November 27, 1998*: Official NSF appointment calendar for the director, courtesy of Kay Risen; Gaffney, phone interview with Colwell, n.d.

111 *"US Navy [w]as a much more impenetrable barrier"*: Kathleen Crane, *Sea Legs: Tales of a Woman Oceanographer* (Boulder, CO: Westview Press, 2003), 293–98; Enrico Bonatti and Kathleen Crane, "Oceanography and Women: Early Challenges," *Oceanography* 25, no. 4 (October 2, 2015).

112 *The US Navy had just completed*: Margo H. Edwards, interview, June 11, 2018, and Edwards and Bernard J. Coakley, "The SCICEX Program:

Arctic Ocean Investigations from a U.S. Navy Nuclear-Powered Submarine," *Arctic Research of the United States* 18 (September 22, 2004).

112 *"important implications for global warming"*: Office of Naval Research, "USS HAWKBILL in Transit to Arctic Ocean for SCICEX 99," press release, March 24, 1999.

112 *"Because when you're out there . . ."*: M. H. Edwards, interview and emails; Edwards and Bernard J. Coakley, "The SCICEX Program Arctic Ocean: Investigations from a U.S. Navy Nuclear-Powered Submarine," n.d.; Dan Steele, "Punching through the Ice Pack," *Professional Mariner*, no. 551 (October/November 2000), www.usshawkbill.com/scicex99.htm; Office of Naval Research, "USS HAWKBILL in Transit to Arctic Ocean for Scicex 99," March 24, 1999; and www.usshawkbill.com. R. R. Colwell, "Polar Connections" (speech delivered at the University Corporation for Atmospheric Research/National Center for Atmospheric Research 40th anniversary and National Science Foundation 50th anniversary celebration, June 19, 2000), NSF Archives, Colwell speeches.

113 Nature *published Edwards's report*: L. Polyak et al., "Ice Shelves in the Pleistocene Arctic Ocean Inferred from Glaciogenic Deep-Sea Bedforms," *Nature* 410, no. 6827 (March 22, 2001): 453–57.

114 *she told her story in* Ice Bound: Jerri Nielsen and Maryanne Vollers, *Ice Bound: A Doctor's Incredible Battle for Survival at the South Pole* (New York: Hyperion, 2001).

116 *John S. Toll, one of the master builders*: Paul Vitello, "John S. Toll Dies at 87; Led Stony Brook University," *New York Times*, July 18, 2011.

117 *men who openly help*: David R. Hekman et al., "Does Diversity-Valuing Behavior Result in Diminished Performance Ratings for Nonwhite and Female Leaders?" *Academy of Management Journal* 60, no. 2 (March 3, 2016).

117 *Revolutions in technology have always fueled*: Jeremy Berg, "Editorial: Revolutionary Technologies," *Science* 361, no. 6405 (August 31, 2018).

117 *The University of Maryland . . . should become a leader in bioscience*: Enrico Moretti, *The New Geography of Jobs* (Boston: Houghton Mifflin Harcourt, 2013).

121 *His wartime science advisor, Vannevar Bush*: Vannevar Bush, *Science: The Endless Frontier: A Report to the President on a Program for Postwar Scientific Research* (Washington, DC: U.S. Government Printing House, 1945).

123 *Supreme Court Justice Ruth Bader Ginsburg*: Adam Liptak, "As Gays Prevail in Supreme Court, Women See Setbacks," *New York Times*, August 4, 2014.

123 *Nancy Pelosi*: Philip Galanes, "A Power Lunch, Times Two," *New York Times*, April 4, 2014.

123 *Denise Faustman*: Stephen Fiedler, "Women in Stem: Q&A with Dr. Denise Faustman of MGH," *SciTech Connect* (blog), Elsevier, March 6, 2014.

124 *the effectiveness of a group*: A. W. Woolley, T. W. Malone, and C. F. Chabris, "Why Some Teams Are Smarter Than Others," *New York Times*, January 16, 2015; A. W. Woolley and T. W. Malone, "What Makes a Team Smarter? More Women," *Harvard Business Review* 89, no. 6 (2011): 32–33.

125 *many students were dropping out*: Joseph Bordogna, interview, June 6, 2013; NSF ed., *The Power of Partnerships: A Guide from the NSF Graduate STEM Fellows in K-12 Education (GK-12) Program* (Washington, DC: American Academy for the Advancement of Science, 2013) provides a guide for creating programs similar to the NSF's GK-12 program.

127 *Ruzena Bajcsy*: Ruzena Bajcsy, interview, September 24, 2018; Ruzena Bajcsy, an oral history conducted by Janet Abbate, Berkeley, CA, July 9, 2002, for the IEEE History Center (IEEE History Center Oral History Program interview #575).

129 *the United States' leadership in mathematics*: National Research Council, *The Mathematical Sciences in 2025* (Washington, DC: National Academies Press, 2013), appendix A; Philippe Tondeur, interview, September 18, 2018, and Tondeur, "NSF: A Wake-Up Call," *Notices of the American Mathematical Society* 52, no. 6 (June/July 2005).

129 *Congress asked me what biocomplexity was*: Rita Colwell, testimony before the Senate Appropriations Committee Subcommittee on VA/HUD and Independent Agencies, May 4, 2000.

130 *Mary E. Clutter*: Mary E. Clutter, interview, August 17, 2013; AD/BBS Circular No. 14, NSF, January 23, 1989; Marcia Clemmit, "Toughest Federal Science Jobs Elude Women," *The Scientist*, October 15, 1990; W. Franklin Harris, "NSF Policy," letter to the editor, *The Scientist*, December 10, 1990.

131 *Joe Bordogna, a former engineering dean*: Joseph Bordogna, interview, Philadelphia, June 6, 2013.

132 *"Show me how to do this," I said*: Margaret Leinen, interview, January 3, 2019.

132 *"one hundred percent of the human race"*: Bordogna, interview.

133 *But how would such a grant work?*: Bordogna, interview; Sue V. Rosser, *Academic Women in STEM Faculty: Views Beyond a Decade After POWRE* (New York: Springer, 2017); Diana Bilimoria and Xiangfen Liang, *Gender Equity in Science and Engineering: Advancing Chang in Higher Education* (New York: Routledge, 2014), 7–14; "2003 Survey of Doctorate Recipients," NSF Division of Science Resources Statistics.

133 *Gingrich and I were in a meeting*: Rita Colwell, interview by Bill Aspray, July 31, 2017, transcript, National Science Foundation Directorate for Computer and Information Science and Engineering.

134 *Bordogna heard Senator John McCain*: Hearing Summary: Senate Committee on Health, Education, Labor and Pensions National Science Foundation Fiscal 2003 Budget Request, June 19, 2002, www.nsf.gov/about/congress/107/hs_061902help.jsp.

134 *In the end, I couldn't double*: From the time I started as director of the NSF until I left in 2004, the NSF's budget increased by 63 percent, from $3.43 billion to $5.589 billion. To calculate the percentage increase, first get the increase in dollars by subtracting ($5,589,000,000 – $3,430,000,000, which equals $2,159,000,000). The dollar increase is then divided by the starting amount ($3,430,000,000) to get 0.629 or 63%.

Below is the NSF's yearly budget and percent increase over the prior year for the years I was director (all figures come from the NSF's National Center for Science and Engineering Statistics):

1998: $3,430,630,000 (start of my term)
1999: $3,676,050,000 / 7.15% increase
2000: $3,912,000,000 / 6.41% increase
2001: $4,430,570,000 / 13.26% increase
2002: $4,823,350,000 / 8.87% increase
2003: $5,323,090,000 / 10.36% increase
2004: $5,588,860,000 / 4.99% increase

We use the 1998 budget as our base year for the increase because, although I did not prepare the 1999 budget, I worked on it, defending it in Congress, to the Office of Management and Budget, and to the White House. Looking through my official NSF appointment diary, I see that after I became director in August 1998, I visited more than a dozen members of Congress to discuss their support for the budget.

chapter seven: The Anthrax Letters

137 *I approached John R. Phillips*: John R. Phillips, interview, Oakton, Virginia, March 20, 2013, and emails.

138 *invited to join the project's advisory board*: I was a member of a task force called MEDEA, which evolved out of a task force organized by Vice President Al Gore and Linda Zall of the CIA, under President Bill Clinton's administration, to explore ways that intelligence satellite data could fill critical gaps in data for environmental scientists. MEDEA was shut down by the Bush administration and resumed by Senator Dianne Feinstein shortly after President Barack Obama's election in 2008; Phillips, interview; Linda Zall, interview, n.d.; John M. Deutch, "The Environment on

the Intelligence Agenda" (speech delivered at the World Affairs Council, Los Angeles, CA, July 25, 1996).

139 *assumed al-Qaeda was following up*: Jacob Weisberg, *The Bush Tragedy* (New York: Random House, 2008), 190–91.

139 *one of the most likely biological agents*: Centers for Disease Control and Prevention, "Anthrax: The Threat," www.CDC.gov/anthrax/bioterrorism/threat.html; David Willman, *The Mirage Man*: *Bruce Ivins, the Anthrax Attacks, and America's Rush to War* (New York: Bantam Books, 2011), 15, 85; paperback with new title, *The Ames Strain: The Mystery Behind America's Most Deadly Bioterror Attack* (Brooklyn, NY: February Books, 2014).

139 *Anthrax infects primarily large grazing animals*: World Health Organization, *Anthrax in Humans and Animals*, 4th ed. (Geneva, Switzerland: World Health, 2008).

140 *stunning early evidence of climate change*: William Broad, "CIA Is Sharing Data with Climate Scientists," *New York Times*, January 4, 2010; Aant Elzinga, ed., *Changing Trends in Antarctic Research* (The Netherlands: Springer, 1993).

141 *I invited Phillips to meet with*: Phillips, interview, Oakton, Virginia, March 20, 2013, and emails.

141 *the FBI employed only two microbiologists*: Scott Decker, interview, January 15, 2015; B. Budowle, S. E. Schutzer, and R. G. Breeze, eds., *Microbial Forensics*, 1st ed. (The Netherlands: Elsevier Academic Press, 2005); R. R. Colwell, "Forward," in *Microbial Forensics*, eds. Budowle et al.; S. A. Morse and B. Budowle, "Microbial Forensics: Application to Bioterrorism Preparedness and Response," *Infectious Disease Clinics of North America* 20, no. 2 (2006): 455–73.

142 *the White House and Congress were getting an education*: George W. Bush, *Decision Points* (New York: Crown Publishing, 2010), 157–58; Leonard A. Cole, *The Anthrax Letters: A Bioterrorism Expert Investigates the Attacks That Shocked America* (New York: Skyhorse Publishing, 2009), 117; A. Scorpio et al., "Anthrax Vaccines: Pasteur to the Present," *Cellular and Molecular Life Sciences* 63, no. 19–20 (October 2006): 2237–48; Weisberg, *The Bush Tragedy*, 190–91; Fred Charatan, "Bayer Cuts Price of Ciprofloxacin After Bush Threatens to Buy Generics," *British Medical Journal (BMJ)* 323, no. 7320 (2008); L. M. Wein, D. L. Craft, and E. H. Kaplan, "Emergency Response to an Anthrax Attack," *PNAS* 100, no. 7 (April 1, 2003).

142 *a lone terrorist could make an anthrax weapon*: G. F. Webb, "A Silent Bomb: The Risk of Anthrax as a Weapon of Mass Destruction," *PNAS* 100, no. 8 (April 15, 2003): 4355–56.

142 B. anthracis *found in one part of the world*: Talima Pearson et al., "Phylogenetic Discovery Bias in *Bacillus anthracis* Using Single-Nucleotide

Polymorphisms from Whole-Genome Sequencing," *PNAS* 101, no. 37 (September 14, 2004): 13536–41; P. Keim and K. L. Smith, "*Bacillus anthracis* Evolution and Epidemiology," in *Anthrax (Current Topics in Microbiology and Immunology)* vol. 271, ed. T. M. Koehler (Berlin: Springer, 2002), 21–32.

143 *Fraser "the undisputed world leader in microbial genomics"*: Martin Enserink and Andrew Lawler, "Research Chiefs Hunt for Details in Proposal for New Department," *Science* 296, no. 5575 (June 14, 2002): 1944–45.

144 *"a single crazy incident"*: Claire M. Fraser, interview, April 4, 2013.

144 *Dr. Larry M. Bush*: L. M. Bush et al., "Index Case of Fatal Inhalational Anthrax Due to Bioterrorism in the United States," *New England Journal of Medicine* 345 (November 29, 2001): 1607–10.

144 *Concerns about* B. anthracis *had circulated recently*: T. V. Inglesby et al., "Anthrax as a Biological Weapon, 2002: Updated Recommendations for Management," *JAMA* 287 (2002): 2236–52; P. Keim et al., "Molecular Investigation of the Aum Shinrikyo Anthrax Release in Kameido, Japan," *Journal of Clinical Microbiology* 39, no. 12 (December 2001): 4566–67; Raymond A. Zilinskas, "The Soviet Biological Warfare Program and Its Uncertain Legacy," *Microbe* 9, no. 5 (2014): 191–97; Matthew Meselson et al., "The Sverdlovsk Anthrax Outbreak of 1979," *Science* 266, no. 5188 (November 18, 1994).

144 *An office worker in New York City*: Bella English, "Struggles Remain for Victim of Anthrax Attack," *Boston Globe*, September 16, 2012.

145 *More anthrax-laden letters arrived*: Willman, *The Mirage Man*, 423, endnote 6; Michael R. Kuhlman, interview, November 18, 2015.

145 *TIGR applied for a grant*: NSF, NSF Archives Award Abstract #0202304 (October 26, 2001).

145 *Homeland Security director Tom Ridge*: "Gov. Ridge, Medical Authorities Discuss Anthrax," press briefing transcript, October 25, 2001, https://georgewbush-whitehouse.archives.gov/news/releases/2001/10/20011025.-4.html.

146 *Paul L. Jackson, a Los Alamos National Laboratory authority*: Paul L. Jackson, interview, September 5, 2014.

146 *Keim . . . and Popovic*: Paul Keim, interview, April 19, 2013; Phillips, interview, March 22, 2013.

148 *Ames had gone to laboratories in only three foreign countries*: Jeanne Guillemin, *American Anthrax: Fear, Crime, and the Investigation of the Nation's Deadliest Bioterror Attack* (New York: Macmillan, 2011), 146–47.

148 *Michael R. Kuhlman*: Kuhlman, interview.

148 *Teresa "Terry" G. Abshire*: Terry Abshire, interviews, June 18 and July 29, 2014.

150 *"We were panicked"*: Richard Cohen, "Our Forgotten Panic," *Washington Post*, July 22, 2014.

151 *A recently disbanded White House task force*: Thomas A. Cebula, interview, April 10, 2013.

151 *Ari Patrinos*: Ari A. N. Patrinos, interview, May 2, 2013.

152 *more than seventeen agencies would know*: Members included the National Science Foundation; the Intelligence Community, notably Phillips's Office of Chief Scientist at the CIA; both the Department of Justice and the FBI; the NIH (the CDC and the National Institute of Allergy and Infectious Diseases, or NIAID); components of what became the Department of Homeland Security and later the department itself; DARPA (Defense Advanced Research Projects Agency); and the Departments of Agriculture and Energy. These were joined later by the National Biodefense Analysis and Countermeasures Center; the FDA; the EPA; the US Army Medical Research Institute of Infectious Diseases (USAMRIID) at Fort Detrick, Frederick, Maryland, which was developing vaccines against anthrax and other pathogens for military personnel; DTRA (Defense Threat Reduction Agency); the Transformational Medical Technologies Initiative, interested in countering biological weapons for the Department of Defense; the Naval Medical Research Center; Edgewood Chemical Biological Center; and the NSA.

153 *SCIF*: Keim, interview; Ronald. A. Walters, interview, March 15, 2013; Daniel Drell, interview, May 3, 2013.

153 *Giovanni at the NIH offered to fund*: Maria Giovanni, email, December 15, 2017.

153 *Bruce E. Ivins . . . "You can't really tell one particular strain"*: Willman, *The Mirage Man*, 138–39.

154 "Star Wars *stuff"*: National Research Council, *Approaches Used During the FBI's Investigation of the 2001 Anthrax Letters* (Washington, DC: National Academies Press, 2011).

154 *Budowle told his bosses at the agency*: Bruce Budowle, interview, October 2014.

155 *An assembly line*: Keim, interview; Timothy D. Read, interview, April 16, 2013; Jacques Ravel, interview, March 25, 2013; Steven L. Salzberg, interview, January 2015, and email, July 23, 2019; Mihai Pop, interview, July 24, 2019; Adam Phillippy, interview, July 2019.

157 *Ravel and several FBI agents with PhDs*: Scott T. Stanley, interview, January 15, 2015; R. Scott Decker, "Amerithrax: The Realization of Biological Terrorism," *The Grapevine*, October 2014.

157 *In September 2007*: Willman, *The Mirage Man*, 252–53; Ravel, interview, March 10, 2015.

157 *"five years of work had contributed something"*: Ravel, interview.

157 *soon the FBI could say*: US Department of Justice, *The Science: Anthrax Press Briefing* (August 18, 2008); US Department of Justice, *Amerithrax Investigative Summary* (February 19, 2010), 25–26; Willman, *The Mirage Man*, 255.

159 *"likely prevented a mass shooting"*: Gregory Saathoff, *Amerithrax Case: The Report of the Expert Behavioral Analysis Panel* (Montreal: Libly, August 20, 2010), 8, 11.

159 *the FBI published the 92-page* Amerithrax Investigative Summary: US Department of Justice, *Amerithrax Investigative Summary*, 1.

159 *Today, whole-genome sequencing methods*: W. F. Fricke, D. A. Rasko, and J. Ravel, "The Role of Genomics in the Identification, Prediction, and Prevention of Biological Threats," *PLOS Biology* 7, no. 10 (October 2009); S. J. Joseph and T. D. Read, "Bacterial Population Genomics and Infectious Disease Diagnostics," *Trends in Biotechnology* 28, no. 12 (December 2010): 611–18.

chapter eight: **From Old Boys' Clubs to Young Boys' Clubs to Philanthropists**

162 *Half of our PhD scientists are already migrating*: Katie Langin, "Private Sector Nears Rank of Top PhD Employer," *Science* 363, no. 6432 (March 15, 2019): 1135.

162 *producing too many science PhDs*: Gina Kolata, "So Many Research Scientists, So Few Openings as Professors," *New York Times*, July 14, 2016.

163 *Catalyst*: "The Promise of Future Leadership: Highly Talented Employees in the Pipeline," Catalyst, February 2001.

163 *only twenty-four women were CEOs of the . . . S&P 500*: C. C. Miller, K. Quealy, and M. Sanger-Katz, "The Top Jobs Where Women Are Outnumbered by Men Named John," *New York Times*, April 24, 2018; Shawn Tully, "Outnumbered by Jeffreys," *Fortune*, June 28, 2019.

163 *Christine Lagarde*: David Segal and Amie Tsang, "Call in the Woman: Lagarde to Steer Europe in Rough Economic Seas," *New York Times*, July 3, 2019.

164 *I boldly started my own company*: Rita Colwell interviewed Manoj Dadlani on December 4, 2019, and verbally relayed the section on CosmosID; Dadlani approved/agreed. On December 11, 2019, Colwell interviewed Bruce Grant and he acknowledged that Dadlani had agreed with the section.

164 *didn't know where to turn for advice*: Robbie Melton, interview, March 5, 2019.

165 *men overestimate their abilities by as much as 30 percent*: Jena McGregor, "Yet Another Explanation for Why Fewer Women Make It to the Top,"

Washington Post, November 29, 2011; Kay and Claire Shipman, "The Confidence Gap," *The Atlantic*, May 2014.

165 *Ernesto Reuben*: Ernesto Reuben et al., "The Emergence of Male Leadership in Competitive Environments," *Journal of Economic Behavior and Organization* 83, no. 1 (2012): 111–17.

166 *knowing how to negotiate*: Linda Babcock and Sara Laschever, *Women Don't Ask: Negotiation and the Gender Divide* (Princeton, NJ: Princeton University Press, 2003).

166 *I read Katty Kay and Claire Shipman's article*: Kay and Shipman, "The Confidence Gap."

167 *their companies will make more money*: Credit Suisse Research Institute, *Gender Diversity and Corporate Performance*, July 31, 2012; *Ernst & Young*: Jenny Anderson, "Huge Study Finds that Companies with More Women Leaders Are More Profitable," Quartz, February 8, 2016; *McKinsey*: V. Hunt et al., "Delivering through Diversity," McKinsey & Company, January 2018; and Lily Trager, "Why Gender Diversity May Lead to Better Returns for Investors," Morgan Stanley, March 7, 2019.

167 *International Monetary Fund*: Lone Christiansen et al., "Gender Diversity in Senior Positions and Firm Performance: Evidence from Europe," IMF, March 2016, quoted in Emily Chang, *Brotopia: Breaking Up the Boys' Club of Silicon Valley* (New York: Penguin, 2019), 254.

167 *It's not women per se who make profits*: Chang, *Brotopia*, 254.

167 *turnover in executive boardrooms*: Jeff Green, "Women May Not Reach Boardroom Parity for 40 Years, GAO says," *Bloomberg*, January 4, 2016.

167 *Approximately 25 percent of the growth in this nation's GDP*: C. T. Hsieh et al., "The Allocation of Talent and U.S. Economic Growth," *Econometrica* 87, no. 5 (September 2019); Paul Gompers and Silpa Kovvali, "The Other Diversity Dividend," *Harvard Business Review*, July-August 2018; Emily Chasan, "The Last All-Male Board in the S&P 500 Finally Added a Woman," Bloomberg, July 24, 2019; Yaron G. Nili, "Beyond the Numbers: Substantive Gender Diversity in Boardrooms," *Indiana Law Journal* 94, no. 1 (2019).

168 *two big studies of corporate America and women in STEMM*: S. A. Hewlett, B. C. Luce, and L. J. Servon, *The Athena Factor: Reversing the Brain Drain in Science, Engineering, and Technology* (Cambridge, MA: Harvard Business Review, May 2008); Caroline Simard and Andrea Davies Henderson, *Climbing the Technical Ladder: Obstacles and Solutions for Mid-Level Women in Technology* (Palo Alto and Stanford, CA: Anita Borg Institute for Women and Technology in collaboration with the Clayman Institute for Gender Research, 2008); Kathleen Melymuka, "Why Women Quit Technology," *Computerworld*, June 16, 2008.

168 *frat-house atmosphere*: Dan Lyons, "Jerks and the Start-Ups They Ruin," *New York Times*, April 1, 2017.

168 *Employees at other early start-ups*: Josh Harkinson, "Welcome Back to Silicon Valley's Biggest Sausage Fest," *Mother Jones*, September 9, 2014; Lester Haines, "Apple Squashes Wobbly Jub App," *The Register*, February 19, 2010, https://www.theregister.co.uk/2010/02/19/app_squashed/; Betsy Morais, "The Unfunniest Joke in Technology," *The New Yorker*, September 9, 2013; Claire Cain Miller, "Technology's Man Problem," *New York Times*, April 5, 2014; Susan Fowler, "Reflecting on One Very, Very Strange Year at Uber," *Susan Fowler* (blog), February 19, 2017, https://www.susanjfowler.com/blog/2017/2/19/reflecting-on-one-very-strange-year-at-uber; Fowler, "I Wrote the Uber Memo. This is How to End Sexual Harassment," *New York Times*, April 12, 2018; Yoree Koh, "Uber's Party Is Over: New Curbs on Alcohol, Office Flings," *Wall Street Journal*, June 13, 2017; and Chang, *Brotopia,* 106–35.

169 *Peter Thiel*: Peter Thiel, "The Education of a Libertarian," Cato Unbound, April 13, 2009, https://www.cato-unbound.org/2009/04/13/peter-thiel/education-libertarian.

169 *As a result of toxic work environments*: Chang, *Brotopia*, 14–15.

169 *ageism operates "all the time"*: Amy Millman, interview, May 1, 2019.

169 *Venture capitalists perform a public service*: Alison Wood Brooks et al., "Investors Prefer Entrepreneurial Ventures Pitched by Attractive Men," *PNAS* 111, no. 2 (March 25, 2014).

169 *women founders received less than 3 percent of all venture capital funding*: Candida Brush, interview, April 16, 2019; The Diana Project, Babson College, Wellesley, MA, https://www.babson.edu/academics/centers-and-institutes/center-for-womens-entrepreneurial-leadership/thought-leadership/diana-international/diana-project/.

169 *"staggeringly" homogeneous*: Gompers and Kovvali, "The Other Diversity Dividend," 72–77.

170 *Entrepreneurs from one generation of tech companies invest*: Margaret O'Mara, "Silicon Valley's Old Money," *New York Times*, March 31, 2019.

170 *prefer to invest in men*: Brooks et al., "Investors Prefer Entrepreneurial Ventures Pitched by Attractive Men," 4427–31.

170 *if a senior partner has a daughter*: Paul A. Gompers and Sophie Q. Wang, "And the Children Shall Lead: Gender Diversity and Performance in Venture Capital" (National Bureau of Economic Research Working Paper No. 23454, May 2017).

170 *Carol A. Nacy*: Carol A. Nacy, interview, April 11, 2018.

171 *universities allow investors*: Nancy Hopkins, interview, January 21, 2018;

Hopkins, "Lost in the Biology-to-Biotech Pipeline: A Tale of 2 Leaks" (Rosalind Franklin Society board meeting, December 17, 2014).

171 *Students are especially vulnerable*: Hopkins, interview with Colwell and McGrayne, March 15, 2018.

171 *science philanthropy*: Fiona Murray, "Evaluating the Role of Science Philanthropy in American Research Universities," in *Innovation Policy and the Economy*, vol. 13, eds. Josh Lerner and Scott Stern (Chicago: University of Chicago Press, 2013): 23–60.

172 *Marc Kastner*: Kate Zernike, "Gains, and Drawbacks, for Female Professors," *New York Times*, March 21, 2011.

172 *donations . . . do not include overhead*: Marc Kastner, interview, August 6, 2019.

173 *professionalizing the master of science degree*: R. R. Colwell, "Professional Science Master's Programs Merit Wider Support," *Science* 323, no. 5922 (March 27, 2009).

173 *Fortunately, some women aren't waiting*: Millman, interview; Brush, interview.

175 *Gulf of Mexico Research Initiative*: www.gulfresearchinitiative.org.

176 *It took a lot of fine print*: GoMRI, "About the Gulf of Mexico Research Initiative Research Board," https://gulfresearchinitiative.org/gri-research-board/.

177 *Thanks to GoMRI scientists, we now know more*: Charles "Chuck" Miller, interview, September 5, 2018; Claire B. Paris et al., "BP Gulf Science Data Reveals Ineffectual Subsea Dispersant Injection for the Macondo Blowout," *Frontiers in Marine Science*, October 30, 2018.

178 *Soderlund warned me*: Karl Soderlund, interview, September 12, 2018.

179 *World Health Organization and UNICEF say that one in three people on Earth . . . lack safe water*: "Drinking-water" (World Health Organization fact sheet), www.who.int/en/news-room/fact-sheets/detail/drinking-water, accessed November 15, 2019.

chapter nine: **It's Not Personal—It's the System**

181 *Girls have earned higher grades*: Daniel Voyer and Susan D. Voyer, "Gender Differences in Scholastic Achievement: A Meta-Analysis," *Psychological Bulletin* 140, no. 4 (April 29, 2014).

181 *women don't underestimate their own abilities as much as men* overestimate *theirs*: Mary A. Lundeberg, Paul W. Fox, and Judith Punćochař, "Highly Confident but Wrong: Gender Differences and Similarities in Confidence Judgments," *Journal of Educational Psychology* 86, no. 1 (1994).

182 *enormous cost of discrimination against women*: Nicole Smith, interviews, April 23, May 7, June 10, 2019, and email February 11, 2020.

183 *Social scientists tell us that implicit bias*: Claude M. Steele, *Whistling Vivaldi: How Stereotypes Affect Us and What We Can Do* (New York: W. W. Norton, 2010); and Mahzarin R. Banaji and Anthony G. Greenwald, *Blindspot: Hidden Biases of Good People* (New York: Penguin Random House, 2013).

183 *Implicit bias can be overcome with rational, careful deliberation*: Jay Van Bavel, interview, August 20, 2019.

184 *auditioning new players behind a curtain*: Geoff Edgers, "Elizabeth Rowe Has Sued the BSO: Her Case Could Change How Orchestras Pay Men and Women," *Boston Globe*, December 11, 2018; Malcolm Bay, "BSO Flutist Settles Equal-Pay Lawsuit with Orchestra," *Boston Globe*, February 14, 2019.

184 *Jennifer T. Chayes*: Jennifer T. Chayes, interview, January 28, 2019.

184 *Jo Handelsman*: Jo Handelsman, interview, February 26, 2013; C. A. Moss-Racusin et al., "Science Faculty's Subtle Gender Biases Favor Male Students," *PNAS* 109, no. 41 (October 9, 2012).

185 *Virginia Valian*: Virginia Valian, *Why So Slow? The Advancement of Women* (Cambridge, MA: MIT Press, 1998); and Natalie Angier, "Exploring the Gender Gap and the Absence of Equality," *New York Times*, August 25, 1998.

185 *Nobel Prize winner Elizabeth Blackburn*: Mallory Pickett, "I Want What My Male Colleague Has, and That Will Cost a Few Million Dollars," *New York Times Magazine*, April 18, 2019.

186 *Jason Sheltzer*: J. M. Sheltzer and J. C. Smith, "Elite Male Faculty in the Life Sciences Employ Fewer Women," *PNAS* 111, no. 28 (July 15, 2014).

186 *Sir Richard T. "Tim" Hunt*: Rebecca Ratcliffe et al., "Nobel Scientist Tim Hunt: Female Scientists Cause Trouble for Men in Labs," *Guardian* (UK), June 10, 2015.

186 *Science scandals featured male celebrity professors*: Tamar Lewin, "Yale Medical School Removes Doctor after Sexual Harassment Finding," *New York Times*, November 14, 2014; Lewin, "Seven Allege Harassment by Yale Doctor at Clinic," *New York Times*, April 15, 2015; Alexandra Witze, "Astronomy Roiled Again by Sexual-Harassment Allegations," *Nature*, January 13, 2016; Jeffrey Mervis, "Caltech Suspends Professor for Harassment," *Science*, January 1, 2016; Dennis Overbye, "Geoffrey Marcy to Resign from Berkeley Astronomy Department," *New York Times*, October 14, 2015; Amy Harmon, "Chicago Professor Resigns amid Sexual Misconduct Investigation," *New York Times*, February 3, 2016; Michael Balter, "The Sexual Misconduct Case That Has Roiled Anthropology," *Science*, February 9, 2016; Katherine Long, "UW Researcher Michael Katze Fired After Sexual-Harassment Investigation," *Seattle Times*, August 3, 2017;

Anemona Hartocollis, "Dartmouth Reaches $14 Million Settlement in Sexual Abuse Lawsuit," *New York Times*, August 6, 2019.

187 *1,700 male undergraduate biology students*: Daniel Z. Grunspan et al., "Male Biology Students Consistently Underestimate Female Peers, Study Finds," *PLOS ONE* 11, no. 2 (February 10, 2016).

187 *women's papers are cited more*: Elizabeth Culotta, "Study: Male Scientists Publish More, Women Cited More," *The Scientist* (July 1993); Rachel Pells, "Male Authors Tend to Cite Male Authors More Than Female Authors," *Inside Higher Education*, August 16, 2018.

187 *Men think computer code written by women is better*: Tia Ghose, "Female Coders Get Less Respect When Their Gender Shows," *Washington Post*, February 23, 2016.

187 *Letters of recommendation for women are shorter*: Kuheli Dutt et al., "Gender Differences in Recommendation Letters for Postdoctoral Fellowship in Geoscience," *Nature Geoscience* 9, no. 11 (October 3, 2016); Sarah-Jane Leslie et al., "Expectations of Brilliance Underlie Gender Distributions Across Academic Disciplines," *Science* 347, no. 6219 (January 16, 2015); Rachel Bernstein, "Belief That Some Fields Require 'Brilliance' May Keep Women Out," *Science*, January 15, 2015.

187 *must publish three more papers*: Lawrence K. Altman, "Swedish Study Finds Sex Bias In Getting Science Jobs," *New York Times*, May 22, 1997.

187 *Men talk shop with their male STEMM colleagues*: S. E. Holloran et al., "Talking Shop and Shooting the Breeze: A Study of Workplace Conversation and Job Disengagement among STEM Faculty," *Social Psychological and Personality Science* 2, no. 1 (2011).

187 *In some fields, like economics, women get zero recognition*: Heather Sarsons, "Recognition for Group Work: Gender Differences in Academia," *American Economic Review* 107, no. 5 (May 2017).

187 *Two women who submitted an article to . . .* PLOS ONE: Holly Else, "'Sexist' Peer Review Causes Storm Online," *Times Higher Education*, April 3, 2015; Damian Pattinson, "PLOS ONE Update on Peer Review Process," *EveryONE* (blog), *PLOS ONE*, https://blogs.plos.org/everyone/2015/05/01/plos-one-update-peer-review-investigation/.

188 *Female and other minority executives who promote diversity are penalized*: David R. Hekman, "Does Diversity-Valuing Behavior Result in Diminished Performance Ratings for Nonwhite and Female Leaders?" *Academy of Management Review*, March 3, 2016.

188 *Male faculty in STEMM are more reluctant than women to believe research about gender bias*: Ian M. Handley et al., "Quality of Evidence Revealing Subtle Gender Biases in Science Is in the Eye of the Beholder," *PNAS* 112, no. 43 (October 27, 2017).

188 *training can actually increase the problem*: Virginia Gewin, "Why Some Anti-Bias Training Misses the Mark," *Nature*, April 22, 2019.

188 *suits against the elite Salk Institute for Biological Studies*: Mallory Pickett, "I Want What My Male Colleague Has, and That Will Cost a Few Million Dollars," *New York Times Magazine*, April 18, 2019; Meredith Wadman, "Salk Institute Hit with Discrimination Lawsuit by Third Female Scientist," *Science*, July 20, 2017.

188 *the NIH gives bigger grants to men*: Andrew Jacobs, "Another Obstacle for Women in Science: Men Get More Federal Grant Money," *New York Times*, March 5, 2019.

188 *speaking invitations to women who deserve them*: Arturo Casadevall and Jo Handelsman, "The Presence of Female Conveners Correlates with a Higher Proportion of Female Speakers at Scientific Symposia," *mBio* 5, no. 1 (January/February 2014); Arturo Casadevall, "Achieving Speaker Gender Equity at the American Society for Microbiology General Meeting," *mBio* 6, no. 4 (August 4, 2015).

189 *Greg Martin calculated the odds of a panel being "randomly" all male*: Lauren Bacon, "The Odds That a Panel Would 'Randomly' Be All Men Are Astronomical," *The Atlantic*, October 20, 2015.

189 *the university ranks eighth in the United States for graduating black PhDs*: Crystal Brown (chief communications officer, University of Maryland), email, September 26, 2014, and interview with email, October 27, 2019; Natifia Mullings (director of communications, University of Maryland), interview and email, October 27, 2019.

191 *a remarkable group for the study*: National Academies of Sciences, Engineering, and Medicine, *Sexual Harassment of Women: Climate, Culture, and Consequences in Academic Sciences, Engineering, and Medicine* (Washington, DC: National Academies Press, 2018), v.

191 *The report had clear findings*: National Academies of Sciences, Engineering, Medicine, *Sexual Harassment of Women*.

192 *Months before the report was published*: France Córdova, email, April 16, 2020; France Córdova, "Leadership to change a culture of sexual harassment," Science, March 27, 2020; National Science Foundation, *Federal Register* 83, 47940 (2018).

192 *now requires institutions to report*: Alexandra Witze, "Top U.S. Science Agency Unveils Hotly Anticipated Harassment Policy," *Nature* 561, no. 7724 (2018); Megan Thielking, "It's Time for Systemic Change," STAT, September 20, 2018; Meredith Wadman, "In Lopsided Vote, U.S. Science Academy Backs Move to Eject Sexual Harassers," *Science*, April 30, 2019.

193 *The American Geophysical Union recently defined harassment as a form of*

scientific misconduct: Maggie Kuo, "Scientific Society Defines Sexual Harassment as Scientific Misconduct," *Science*, September 20, 2017.

193 *NIH director Francis Collins apologized for the agency's past failure*: Jocelyn Kaiser, "National Institutes of Health Apologizes for Lack of Action on Sexual Harassers," *Science*, February 28, 2019; and Lenny Bernstein, "NIH Director Will No Longer Speak on All-Male Science Panels," *Washington Post*, June 12, 2019.

chapter ten: **We Can Do It**

194 *The best of 100 percent of the population*: Shirley Malcom, interview, April 12, 2019; Nia-Malika Henderson, "White Men Are 31 Percent of the American Population. They Hold 65 Percent of All Elected Offices," *Washington Post*, October 8, 2014; National Academy of Sciences, *Can Earth's and Society's Systems Meet the Needs of 10 Billion People?* (Washington, DC: National Academies Press, 2013).

195 *"Don't take 'no' for an answer"*: Florence Haseltine, interview, September 25, 2016.

196 *listen to Crystal N. Johnson*: Crystal N. Johnson, speech at American Society for Microbiology meeting, New Orleans, Louisiana, May 2015.

196 "*Teach girls to stand up for themselves*": Shirley M. Tilghman, interview, May 21, 2019.

197 *Bring scientists and engineers . . . into a school*: NSF ed., *The Power of Partnerships: A Guide from the NSF Graduate STEM Fellows in K-12 Education (GK-12) Program* (Washington, DC: American Association for the Advancement of Science, 2013).

197 *Encourage girls*: Lin Bian, Sarah-Jane Leslie, and Andrei Cimpian, "Gender Stereotypes About Intellectual Ability Emerge Early and Influence Children's Interests," *Science* 355, no. 6323 (January 27, 2017).

197 *Start encouraging girls' confidence*: Google, "Women Who Choose Computer Science—What Really Matters: The Critical Role of Encouragement and Exposure," May 26, 2014, https://edu.google.com/pdfs/women-who-choose-what-really.pdf.

198 *Study non-STEMM disciplines, too*: Satyan Linus Devadoss, "A Math Problem around Pi Day," *Washington Post*, March 17, 2018.

199 *Stay in school*: Peter Schmidt, "Men's Share of College Enrollments Will Continue to Dwindle, Federal Report Says," *Chronicle of Higher Education*, May 27, 2010.

201 *science junkies*: Tamar Barkay, interview, June 14, 2016.

201 *Go on field trips, but gather information before you go*: K. B. H. Clancy et al., "Survey of Academic Field Experiences (SAFE): Trainees Report Harassment and Assault," *PLOS ONE* 9, no. 7 (July 16, 2014).

202 *Physicist Fay Ajzenberg-Selove*: Fay Ajzenberg-Selove, *A Matter of Choices: Memoirs of a Female Physicist* (New Brunswick, NJ: Rutgers University Press, 1994).

202 *steps to take before choosing a PhD advisor*: Constance L. Cepko, interview, February 7, 2018.

203 *"Science is different from other professions"*: Marcia McNutt, interview, July 21, 2019.

203 *Academic science is more family friendly*: Cepko, interview.

204 *Don't agonize*: Ian Sample, "Nobel Winner Declares Boycott of Top Science Journals," *Guardian* (UK), December 9, 2013.

204 *If you're isolated*: A. J. Cox et al., "For Female Physicists, Peer Mentoring Can Combat Isolation," *Physics Today*, October 18, 2014.

204 *Men are four times*: Babcock and Laschever, *Women Don't Ask*.

205 *Ask people to nominate you for prizes*: Florence Haseltine, interview, September 25, 2016.

205 *Make a habit of empowering other women*: Haseltine, interview, September 25, 2016.

205 *"Let the men do it for a while"*: Alice Huang, speaking at the Rosalind Franklin Society annual meeting, December 17, 2014.

206 *"I have a lovely set of colleagues, but"*: Confidential interview, August 30, 2016.

208 *Excluding women from science and technology can be highly detrimental to the country*: Marie Hicks, interview, March 27, 2018, and Hicks, *Programmed Inequality: How Britain Discarded Women Technologists and Lost Its Edge in Computing* (Cambridge, MA: MIT Press, 2017).

209 *does not believe in evolution*: Megan Brenan, "40% of Americans Believe in Creationism," *Gallup News*, July 26, 2019.

209 *The most important issue a woman scientist can speak up about is childcare*: Helen Shen, "Inequality Quantified: Mind the Gender Gap," *Nature*, March 6, 2013; Claire Cain Miller, "The Gender Pay Gap Is Largely Because of Motherhood," *New York Times*, May 13, 2017, and "The 10-Year Baby Window That Is the Key to the Women's Pay Gap," *New York Times*, April 9, 2018; Holly Else, "Nearly Half of US Female Scientists Leave Full-Time Science after First Child," *Nature*, February 19, 2019; Katha Pollitt, "Day Care for All," *New York Times*, February 9, 2019.

210 *the proportion of all grant funds awarded to scientists under the age of thirty-six*

fell: B. Alberts and V. Narayanamurti, "Two Threats to U.S. Science," *Science* 364, no. 6441 (May 17, 2019); John West, interviews, March 27 and 30, 2018.

210 *Amy Millman, founder of Springboard Enterprises*: Amy Millman, interview, May 1, 2019; Jeffrey Mervis, "Top Ph.D. Feeder Schools Are Now Chinese," *Science* 321, no. 5886 (July 11, 2008).

211 *let those we educate in the US*: Alberts and Narayanamurti, "Two Threats to U.S. Science."

211 *Hold scientists who won't mentor women accountable*: Ben A. Barres, emails, September 11, 16, 18, and 19, 2016.

212 *Male faculty . . . use taxpayer-funded research to start all-male tech companies*: Nancy Hopkins, interview with Colwell and McGrayne, March 15, 2018.

212 *Institutionalize your reforms*: John West, interviews.

213 *"civility and respect toward all"*: Ajzenberg-Selove, *A Matter of Choices*, 4.

213 *sexual harassment training can boomerang*: Michelle M. Duguid and Melissa C. Thomas-Hunt, "Condoning Stereotyping: How Awareness of Stereotyping Prevalence Impacts Expression of Stereotypes," *Journal of Applied Psychology* 100, no. 2 (March 2015): 343–59.

214 *scientific organizations have banded together to combat sexual harassment*: Becky Ham, "Societies Take a Stand Against Sexual Harassment with New Initiative," *Science* 363, no. 6434 (March 29, 2019).

214 *Peter Medawar first wrote*: Peter Medawar, *Advice to a Young Scientist* (New York: HarperCollins Children's Books, 1979).

Index

Page numbers beginning with 225 refer to notes.